Taming Self-Interference: Engineering High Isolation for STAR Systems

Sana

First Printing, 2024

Contents

Chapter 1

Introduction

1.1 Background and motivation for this study

As the demand for frequency spectrum increases with new technological advancements in radar and communications, the frequency spectrum is becoming increasingly limited and overcrowded. This is driving research in simultaneous transmit and receive (STAR) systems as there is a growing need for systems with higher spectral efficiency. Conventional systems operate in half-duplex mode, either transmitting and receiving at different times or over different frequency bands. STAR systems essentially operate in full-duplex mode, which can cut spectrum needs in half, thereby doubling the capacity within a given frequency band [1–3].

For a STAR system to be effective, high isolation between transmit and receive channels is required to avoid self-interference. A lack of isolation will result in a significant reduction in receiver sensitivity and dynamic range, reducing its ability to adequately detect incoming signals. This co-channel, self-interference between transmit and receive means that simultaneous transmission and reception (STAR) of signals on the same frequency is an engineering challenge when the transmit and receive channels are co-located.

Transmit-receive isolation is subject to various coupling paths that degrade the system isolation by the transmit signal leaking into the receiver. The direct path between the transmitter and receiver is the dominant source of interference when the transmit and receive antennas are in close proximity to each other. This is what is referred to as self-interference. In addition to this there is also stationary and non-stationary multipath caused by reflections off objects that are stationary and moving around the antennas respectively. These coupling paths are illustrated in Figure 1.1. The combination of all of these produce a multipath environment that can affect transmit-receive isolation. There are also non-linearities and noise which creep in from the hardware in the transmit channel and cause harmonics and intermodulation distortion [4]. The self-interference signal is therefore composed of the transmit signal as well as reflections, non-linearities and noise from the transmit channel.

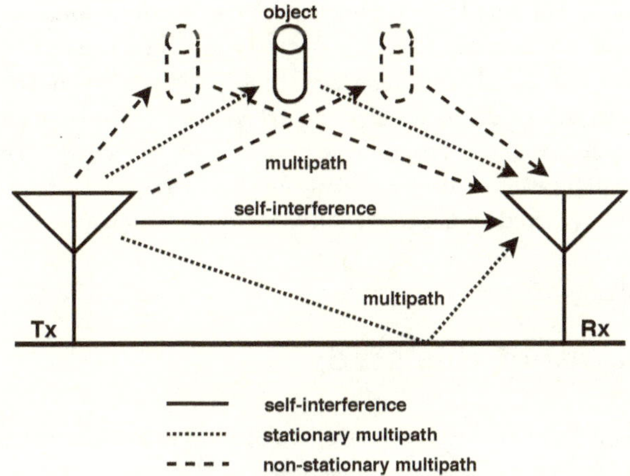

Fig. 1.1. Various coupling paths that degrade system isolation.

There have always been restrictions with the concept of STAR due to insufficient self-interference cancellation, which has been limited by hardware capability. But due to advancements in Radio Frequency (RF) receiver and antenna hardware over the years, better self-interference cancellation has become obtainable and STAR technologies are becoming more and more realisable. STAR systems have applications in radar and communications for military, aerospace and commercial systems. The most common applications are continuous wave (CW) radar and full duplex communication systems.

Pulsed radar systems transmit a signal and then switch off the transmitter in order to receive return echoes. But the problem with that type of radar is that it typically cannot receive signals while it is transmitting due to the powerful transmit signal saturating the receiver. In radar theory, the time that the radar cannot receive signals translates to a particular blind range within which the radar cannot detect any targets. In order to overcome this blind range, techniques such as using different bursts with different PRFs and centre frequencies are used. However, there is still an eclipsing loss that occurs.

In contrast, continuous wave radar systems transmit and receive simultaneously. But self-interference (often referred to by various other labels in radar literature including transmitter leakage, feed-through and direct signal interference) is the major challenge in the design of CW radars. Early CW radars used path loss (bi-static) or circulators (mono-static) to isolate the transmit and receive signals. However, the levels of isolation achieved from circulators imposed a strong limit on the transmit power of the radars (higher transmit power equals stronger self-interference) which in turn limited the radars to short range detection. To overcome this transmit power limitation and increase the range of a CW radar, better self-interference cancellation is required to improve the isolation between transmit and receive.

In the communications field, STAR is commonly referred to as full-duplex. Early application of full-duplex to wireless communications was in the context of relaying where repeaters would receive, amplify and re-transmit the same signal [3]. In recent years the demand for data services has increased drastically and many researchers have looked into the feasibility of full duplex systems which can aid in the growing demands of consumers. The advances in full duplex wireless communications is paving the way for the emerging fifth generation (5G) wireless technologies with the potential for increased link capacity, spectrum virtualisation and improved relaying capabilities.

1.2 Objectives of this study

This study focuses on evaluating the theoretical and practical viability of a STAR system with co-located transmit and receive channels. The principle objectives of this study are to:

- Investigate self-interference techniques to enable STAR capability in an RF system with co-located transmit and receive channels
- Develop a STAR demonstrator with narrowband self-interference cancellation at an ISM frequency band
- Establish the effectiveness of the STAR demonstrator in a multipath environment

1.3 Scope and limitations

The scope of the study is to design and construct a STAR demonstrator using co-located transmit and receive channels and self-interference cancellation at a chosen ISM frequency band. The study is in essence a proof-of-concept study and the aim is not to construct the best possible system. Self-interference cancellation is the key to enabling STAR capability and is therefore the focus area of the STAR demonstrator. The objective is to develop a STAR demonstrator that is effective on narrowband signals. Therefore, wideband operation is not within the scope of this study.

The project budget is limited, so expensive components cannot be bought or manufactured. The antennas used for this study are cheap, off-the-shelf commercial antennas because of their low cost and availability. This may not be the most effective solution, but it is sufficient to demonstrate the concepts in this study. Typically, STAR systems make use of automated analogue self-interference cancellation techniques to achieve optimal cancellation, but for the purposes of this study, the scope does not include automation. However, these limitations should not have a major impact on the main objectives of the study as the scope does not require the development of the most effective system.

1.4 Plan of development

This book is structured in the form of 6 chapters. An overview of the contents of each chapter is summarised below.

Chapter 1 introduces the concept of STAR, and the motivation for this study. The project objectives, scope and limitations are discussed in this chapter along with an outline of the book structure and its content.

Chapter 2 reviews the relevant literature on self-interference cancellation for STAR systems and discusses the various self-interference cancellation techniques that are used.

Chapter 3 discusses the various theory concepts relevant to the study. This includes a description of the relevant self-interference cancellation techniques and non-linear theory.

Chapter 4 discusses the design methodology and detailed design of the STAR demonstrator.

Chapter 5 details the various experiments conducted to characterise the demonstrator, presents the measured results obtained thereof and discusses the results in detail with regards to the objectives of the study.

Chapter 6 summarises the work conducted in this study, draws conclusions from the results and discusses recommendations for future work on the topic.

Chapter 2

Literature Review

The aim of this chapter is to review and analyse the relevant literature pertaining to STAR systems in order to develop an understanding of such systems and to further establish the rationale for this study. STAR systems are also commonly referred to as full duplex systems in the communications field. The two terms are used interchangeably in this chapter and refer to the same concept of transmitting and receiving at the same frequency, at the same time.

A comprehensive review of full duplex research is provided in publications by Sabharwal et al. [3], Liu et al. [5] and Zhang et al. [6]. These publications discuss the various benefits, challenges and techniques to enabling full duplex wireless systems and therefore form the basis of this literature review.

2.1 Fundamentals of full-duplex

It is generally not possible for radios to receive and transmit on the same frequency band because of the interference that results [7].

Conventional systems follow the assumption of the above quote that systems must operate in half-duplex mode. In half-duplex mode, systems either transmit and receive at different times (time division duplexing) or transmit and receive on separate frequency bands (frequency division duplexing). However, research in full-duplex is pointing towards this assumption being invalid. However, the quote does highlight the major challenge to enabling full-duplex, which is self-interference [8].

When a wireless system tries to operate in full-duplex mode it transmits a signal, some of which leaks directly into the system's own receiver. This is referred to as self-interference. Because the transmitter is located near the receiver, this self-interference signal can be in the order of up to 100+ dB stronger than the desired signal which is being received. This can mask signals of interest, saturate and even damage the receiver. Hence, self-interference cancellation is a major enabling factor for full-duplex [5,6,8].

Full duplex systems have the potential to increase spectral efficiency of wireless systems by avoiding the need for two separate channels for bi-directional transmission. This also theoretically means that a doubling of data throughput can be achieved. Many researchers have demonstrated the feasibility of achieving a full duplex system and developed or proposed prototypes for wireless fidelity (WiFi) platforms operating at 2.4 GHz with bandwidths of up to 80 MHz [2,4,9–15]. However, no practical systems have yet been able to fully achieve the theoretical doubling of throughput due to the impact of self-interference [6].

2.2 Self-interference cancellation

The goal of self-interference cancellation is to accurately predict and model the self-interference at the receive antenna in order to be able to reduce the self-interference as much as possible [8]. This cancellation is not a linear operation because the transmitted signal undergoes both linear distortion (from attenuation and multipath reflections) and non-linear distortion (such as quantisation noise, phase noise, intermodulation products and harmonics from non-linear hardware devices) [6]. Effectively the transmitted signal becomes a non-linear function of the ideal transmit signal and simply removing the known transmit signal without accounting for non-linear effects will limit the amount of self-interference cancellation attainable [12].

Self-interference cancellation techniques are classified into 3 main categories: passive suppression, analogue cancellation and digital cancellation. These can then be further sub-categorised as being channel-aware (able to adapt to environmental effects) or channel-unaware (unable to adapt to environmental effects) [5].

Passive suppression

Passive suppression aims to reduce the self-interference in the RF propagation domain before it enters the receiver. It involves achieving high isolation between the transmit and receive antennas to reduce the strength of the self-interference. Analogue and digital cancellation methods suffer from performance degradation which limit the amount of cancellation they can achieve. Hence the need for passive suppression which is capable of providing a large amount of self-interference reduction.

In a shared-antenna system which uses a single antenna for transmit and receive, isolation between the transmit and receive channels is achieved by using a circulator. This isolation is limited by the circulator hardware which has a specified port to port isolation. Bharadia et al. [12] used a shared-antenna in their design with the circulator providing only 15 dB of isolation between transmit and receive. However, they still managed to demonstrate feasibility of full duplex using a shared antenna along with other techniques.

The alternative to using a shared-antenna, is to use separate antennas for transmit and receive. With separate antennas there are further techniques that can be applied to achieve much higher isolation than its single antenna counterpart. One can exploit path loss, directional antennas, polarisation diversity, shielding, antenna placement as well as antenna arrays to increase isolation [3, 5, 6].

The separation of the two antennas gives rise to an inherent path loss. The further away the two antennas are, the higher the path loss, as is demonstrated by [9]. One can also exploit surrounding obstacles or use absorptive RF shielding to block the direct path between antennas. Using directional antennas, the antennas can be placed such that their main beams do not overlap or such that their nulls align. Cross polarisation is another method of enhancing isolation which can also be used in both shared and separate antenna systems [16].

Further ways of improving isolation between transmit and receive antennas involve the use of antenna arrays. Individual transmit antennas can be placed such that they produce a near-field null at the position of the receiver (i.e. the wave forms are 180° out of phase at the receiver and cancel destructively) [1, 10, 17, 18]. There has been a lot of active research on antenna arrays for STAR applications over the past few years by researchers from the University of Colorado [19–28]. There is also transmit beamforming which can be used to steer the main beam as well as nulls in particular directions. Beamforming thus has the added quality of being able to reduce both self-interference as well as reflected path interference off obstacles [29–31]. Furthermore, antenna arrays can also be extended to multiple-input multiple-output (MIMO) systems [32].

The drawback to these passive suppression techniques are that they can be hindered by form factor (size and shape), placement sensitivity, device integration and operating environment. But a more important drawback, with the exception of beamforming, is the sensitivity of these techniques to multipath interference which can reduce the amount of interference suppression in the order of 30 dB compared to measurements in an anechoic chamber [33]. It has also been observed that passive suppression techniques which affect transmit and receive patterns can accidentally suppress desired signals as well [3].

Passive suppression techniques, while capable of providing a large amount of self-interference suppression, may not always be feasible. It also does not necessarily provide sufficient self-interference cancellation to enable high integrity reception of signals of interest. Therefore, additional cancellation techniques are used to further suppress self-interference in the analogue and digital domains to reduce the isolation requirement of passive suppression and improve desired signal integrity.

Analogue cancellation

Analogue cancellation may be done at RF or baseband in the receive chain (i.e. before or after down-conversion), but is typically done at RF. It involves trying to cancel the self-interference by subtracting a copy of the estimated interference signal from the received signal at all instances in the analogue domain before the signal is digitised. Analogue cancellation is not a necessity as digital cancellation can be used directly on the received signal. However, the dynamic range and quantisation resolution of the analogue-to-digital converter (ADC) then becomes limiting on the amount of digital cancellation that can be achieved [34]. In order to avoid this, it is vital to reduce the power of the self-interference as much as possible before it reaches the ADC. Therefore, analogue cancellation typically follows after passive suppression techniques.

The self-interference signal can be modelled as an attenuated and delayed version of the transmit signal. The subtraction of the self-interference signal is achieved by first generating a signal that is an amplitude matched and phase inverted (180° phase inversion) copy of the self-interference signal. This copy is typically generated from the known transmit signal. It is then combined with the received signal. Creating this inverted signal can be done in multiple ways. Importantly, the amplitude of the inverted signal needs to match the amplitude of the self-interference signal for maximum effectiveness. This attenuation (or gain adjustment) is a common component of all analogue cancellation systems. The process of actually inverting the signal can be done in multiple ways. The first method is to apply a phase shift, however this only works across a limited bandwidth. At the centre frequency one would achieve perfect cancellation, but on either side of the centre frequency the phase will deviate from 180° and limit the cancellation ability. This bandwidth limitation can be overcome by using a balanced-unbalanced (balun) transformer [11]. Another method is to use delay-line based analogue circuitry to reconstruct the self-interference signal as a weighted linear combination of samples taken around the recreation instant [35].

Theoretically, with a perfect estimation of the self-interference signal, a perfectly inverted copy of this estimation signal could completely cancel the self-interference [11]. But realistically, practical hardware components have imperfections (e.g. a balun does not have a flat frequency response and will invert a signal with different amplitudes at different frequencies) and cannot produce perfect cancellation. The place where the cancellation signal is tapped off from the transmit signal can also make a difference. The closer to the transmit antenna the signal is tapped off the better it models the self-interference signal as non-linearities such as phase noise and amplifier distortion are captured. The transmit signal can also be tapped off in the digital domain, be digitally processed (with the advantage of being able to account for multipath signal interference) and then converted to analogue for the cancellation [5].

Depending on whether the analogue cancellation can respond to changing environments (multipath effects), it can be either adaptive or non-adaptive. Non-adaptive systems use fixed parameters that are manually adjusted upon design or calibration for gain/attenuation, phase and delay to estimate the self-interference and is therefore sensitive to multipath reflections. Adaptive systems on the other hand dynamically adjust the parameters to be able to deal with both direct self-interference as well as multipath reflections. When the environment changes, a non-adaptive analogue cancellation system's performance may degrade as previously optimal parameters may not be suitable to model the current self-interference. The solution is to adaptively tune the parameters in response to changes in the environment so that optimal cancellation can be achieved at all times. Typically gradient descent convergence algorithms are used to find optimal parameters to minimise the self-interference signal power, either in the time or frequency domain [9–11, 32, 35].

Analogue cancellation does not come without its challenges. Non-linearities and hardware capability place a limit on the amount of cancellation that is attainable. Highly accurate and precise hardware is required to achieve more cancellation, but it is costly. There is also the issues of complexity and size associated with achieving optimal cancellation. Essentially, analogue cancellation is a trade-off between hardware cost and self-interference cancellation capability.

Digital cancellation

Digital cancellation techniques aim to cancel the residual self-interference after digitisation of the signal by the ADC. Even after passive suppression and analogue cancellation, the remaining self-interference can still be strong enough to mask desired signals. Digital cancellation techniques include receive beamforming (adaptively weighting individual antenna returns before combining them) and digital cancellation by estimating the residual interference and subtracting it from the baseband received signal. Receive beamforming can be implemented in the analogue domain, but due to circuit complexity and power requirements it is typically done in the digital domain. The major advantage of digital domain techniques is the reduced complexity that comes with the use of digital signal processing techniques [5].

Digital self-interference cancellation exploits knowledge of the interference signal in order to cancel it after digitisation. This can be achieved by first extracting the self-interference signal and then re-modulating it before subtracting it from the received signal, or through coherent self-interference detection which correlates the received signal with a hypothesised self-interference inversion signal based on the transmit signal [10]. In essence, digital cancellation involves estimating the self-interference and then using the estimation to generate a cancellation signal from the known transmit signal that can be subtracted from the receive signal. It is important to capture both linear and non-linear components in the self-interference estimation to maximise the capability of digital cancellation [6, 11]. As with analogue cancellation, making the process adaptive makes it capable of dealing with changing environment effects. In this case digital cancellation can be seen as digital domain adaptive filtering. A minimisation algorithm can be used to update filter weights and converge to optimal parameters which produce the best cancellation.

One of the practical issues with digital cancellation is that the amount of achievable digital cancellation depends on the efficiency of the preceding self-interference cancellation techniques when they are cascaded (used in conjunction with each other). The amount of cancellation each technique is capable of providing does not add linearly together when cascaded. There are also limitations which stem from practical imperfections and estimation errors. It is therefore important to find the balance between the different techniques in terms of the cost and complexity thereof vs performance to get the best overall solution [6].

Chapter 3

Theory

This chapter discusses relevant theory concepts related to STAR systems. The focus is on the various methods used to enhance transmit-receive isolation in a STAR system (i.e. reduce transmit-receive co-channel self-interference) as well as the basic non-linear theory which arises in RF systems.

A number of self-interference cancellation techniques have been experimented with for STAR systems. High transmit-receive isolation can be achieved through various different techniques in the propagation, analogue RF and digital domains. These cancellation techniques are commonly categorised as passive suppression, analogue cancellation and digital cancellation techniques.

3.1 Passive suppression techniques

Passive suppression techniques are the first line of defence against self-interference. The goal of passive suppression techniques are to isolate the transmit channel from the receive channel electromagnetically, i.e. to suppress self-interference before it enters the receiver, to obtain a high isolation transmit-receive antenna pair. Passive suppression techniques are capable of producing significant isolation. However, various constraints, placement sensitivity and form factor can impact the amount of isolation that can be achieved.

3.1.1 Shared vs Dual Antenna topology

The choice between a shared or dual antenna topology depends on the application. The fundamental difference between the two being that a shared topology uses a single antenna for both transmit and receive, while a shared topology uses separate antennas for transmit and receive. This is illustrated in Figure 3.1.

In a shared topology the transmit-receive isolation is achieved using an RF circulator. An RF circulator is a passive device which regulates signal flow. A signal which flows into one port is transmitted only to the next port in rotation and is isolated from the remaining port/s. The amount of transmit-receive isolation that can be achieved from a shared antenna topology is therefore dependent on the port isolation of the circulator.

A dual antenna topology exhibits a natural isolation due to the separation between the two antennas (path loss). The added advantage of a dual antenna topology is that additional passive suppression techniques can be used to design a high isolation transmit-receive antenna pair. This makes a dual antenna topology capable of achieving very high transmit-receive isolation.

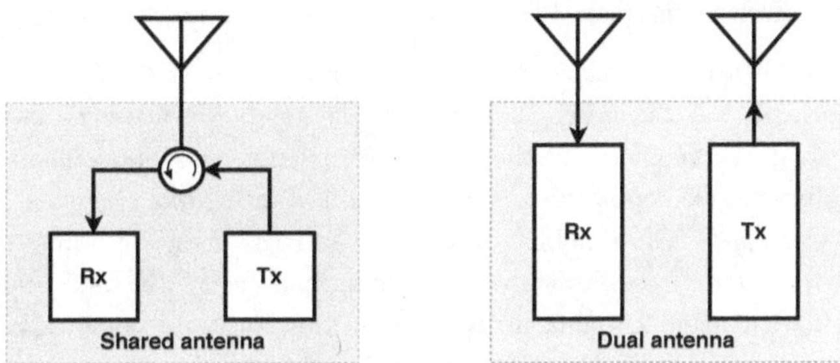

Fig. 3.1. A shared antenna vs dual antenna topology.

3.1.2 Path loss

Path loss (PL) is the attenuation of signal energy between the feed points of two antennas from propagation of the signal in free space with no objects nearby to cause reflections or diffraction. Therefore, by separating the transmit and receive antennas over distance, one can obtain transmit-receive isolation.

The path loss formula is derived from the Friis transmission equation and expresses a loss value assuming the transmit and receive antennas are isotropic, i.e. it takes into account the losses from the spherical spreading of a wave in free space [36]. Path loss is give by the formula

$$PL = (\frac{4\pi d}{\lambda})^2 \tag{3.1}$$

where:
λ is the signal wavelength,
d is the distance between the antennas,
$d \gg \lambda$ such that both antennas are in the far field of each other.

This is defined as the far field path loss as the antennas are required to be in the far field of each other ($d \gg \lambda$). It does not necessarily hold valid if the two antennas are in the near field of each other.

3.1.3 Antenna arrays

Antenna arrays can be used to improve transmit-receive isolation through null synthesis or null steering. Null synthesis involves creating desired nulls in a transmit antenna array radiation pattern and then placing the receive antenna in this null. Null steering is done via transmit beamforming with the goal of adaptively steering nulls in the direction of interference sources. This may involve steering the main beam as well.

The total field of an array is made up of the vector addition of the fields of the individual elements in the array. If the array elements are identical, the array radiation pattern can be controlled using: the geometry of the array, the spacing between elements, the amplitude excitation of individual elements, the phase excitation of individual elements, and the relative pattern of the elements. The number of elements and the array determines the amount of nulls and the degrees of freedom with which nulls can be synthesised and steered. The major drawback with using antenna arrays is that the hardware complexity increases with an increasing number of array elements as each element may require its own amplitude and phase tuning.

Linear arrays are the most basic form of antenna arrays, and are also quite versatile. They have been well studied and documented in many academic textbooks, such as [37]. Linear arrays can produce nulls in the array factor by carefully choosing the element phases and element spacing. There are also various synthesis methods to design arrays with particular characteristics in the radiation pattern, whether it be to shape the main beam, reduce sidelobes or place nulls in desired locations. One such null synthesis method is the Schelkunoff polynomial method. The Schelkunoff polynomial method is a powerful linear antenna array synthesis method to produce nulls in desired locations in a radiation pattern [37]. This synthesis method requires only information on the number of nulls desired, the location of these nulls and the inter-element spacing of the antennas. The number of array elements required and their excitation coefficients can then be derived from this information.

Feeding antenna arrays

An antenna array requires a feed source for each antenna in the array. If there is a large number of array elements this would require a large number of signal sources. The solution is to use an RF power splitter. An RF power splitter is a passive device that coherently divides up a signal into multiple signal channels with equal power. An example of a 4-way splitter is illustrated in Figure 3.2. Passive RF power splitters can also be used as power combiners by applying the input in the opposite direction through the device. RF splitters, like any other practical device, are not perfect and do not provide perfect power coherent division into each output channel and are subject to slight amplitude and phase imbalances.

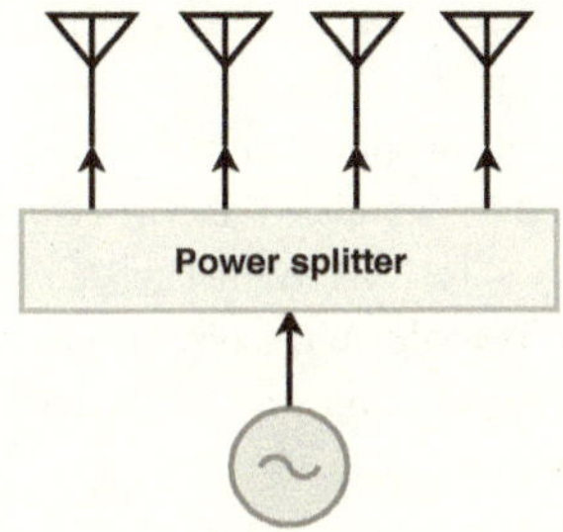

Fig. 3.2. Illustration of a 4 way power splitter feeding an antenna array.

3.2 Analogue cancellation

Analogue cancellation techniques are used to suppress self-interference in the analogue receive chain before the signal is digitised by the ADC. The self-interference signal can be modelled as a scaled and delayed version of the transmit signal. The general principle behind analogue RF cancellation is to take the transmit signal, process the signal in the analogue domain and then subtract it from the received signal to cancel out the interference in the receive channel. Such a system can be adaptive or non-adaptive. A non-adaptive analogue cancellation system with manually calibrated parameters is not capable of dealing with changing environments. An adaptive system, on the other hand, can change its parameters to estimate the interference signal and dynamically adjust to the changing interference signal. The key to achieving good analogue cancellation is to accurately estimate and generate a copy of the interference signal which is to be subtracted from the received signal.

3.3 Digital cancellation techniques

The aim of the digital cancellation layer is to suppress the residual self-interference (as well as multipath clutter) remaining in the digital domain after sampling by applying digital signal processing (DSP) techniques to the received signal. Digital cancellation techniques would typically be adaptive filtering or beamforming. While these techniques can be performed in the analogue domain, their complexity is greatly reduced through implementation in the digital domain. The major disadvantage of digital domain processing is that the ADC's dynamic range limits the amount of self-interference suppression that is possible. Thus in order to effectively use digital domain techniques, a sufficient amount of self-interference must be suppressed before the signal is digitised by the ADC [3].

Adaptive beamforming involves dynamically changing amplitude and phase weights of the transmit antenna array elements to steer radiation pattern nulls in the direction of interference sources while also keeping a null on the receive antenna [4]. This can be done on transmit or receive. Receive beamforming in particular is used in MIMO systems where each antenna's received signal is complex weighted before being collectively summed together. Self-interference is suppressed by adaptively adjusting the antenna weights according to the self-interference condition. While beamforming can be implemented in the analogue domain, digital domain implementations are common because it reduces circuit complexity and power requirements [5].

Adaptive filtering in the digital domain works similarly to analogue RF cancellation. A digital replica of the interference signal is generated and then subtracted from the received signal. [14]

Adaptive filtering involves filtering out the interference (and multipath) from the received signal. Self-interference as well as multipath can be ideally modelled as delayed and scaled versions of the transmit signal which are added to the receive signal. This holds true when the transmitter and receiver are stationary relative to one another and to objects in the environment. Moving objects give rise to dynamic multipath which changes as objects move. The removal of the interference is done through a subtraction operation involving subtracting a reference signal from the received signal. The interference signal is estimated from a reference signal (as the transmit signal) and stored in a matrix. This matrix can contain both delayed versions and well as Doppler shifted versions of the reference signal, thus making it possible to suppress stationary and dynamic multipath components in addition to the self-interference component [38]. Two examples of adaptive filtering algorithms are briefly discussed below.

3.3.1 Extensive Cancellation Algorithm

The Extensive Cancellation Algorithm (ECA) is an adaptive cancellation filter which uses a least squares approach and operates by subtracting weighted, delayed replicas of the reference signal from the surveillance signal, where the weights are adaptively estimated coefficients [39]. Assuming that multipath and echoes are back scattered from the first K range bins, the signal model is exploited by searching for a minimum residual signal power after cancellation of the direct and multipath interference. The ECA algorithm begins from the following minimisation problem:

$$\min_{\alpha} \left\{ ||\mathbf{s_{surv}} - \mathbf{X}\alpha||^2 \right\} \tag{3.2}$$

with

$$\mathbf{X} = \mathbf{B}[\mathbf{\Lambda_{-P}S_{ref}} \cdots \mathbf{\Lambda_{-1}S_{ref}S_{ref}\Lambda_1S_{ref}} \cdots \mathbf{\Lambda_P S_{ref}}] \tag{3.3}$$

where
B is the incidence matrix that selects only the last N rows of the following matrix:

$$\mathbf{B} = \{b_{ij}\}_{i=1,\dots,N;j=1,\dots,N+R-1}, \qquad b_{ij} = \begin{cases} 1, & \text{if } i = j - R + 1 \\ 0, & \text{otherwise} \end{cases} \tag{3.4}$$

$\mathbf{\Lambda_p}$ is a diagonal matrix that applies the phase shift corresponding to the pth doppler bin:

$$\mathbf{\Lambda_p} = \begin{bmatrix} 1 & 0 & \cdots & 0 \\ 0 & e^{j2\pi p} & \cdots & 0 \\ \vdots & \vdots & \ddots & \vdots \\ 0 & 0 & \cdots & e^{j2\pi p(N+R-1)} \end{bmatrix} \tag{3.5}$$

$\mathbf{S_{ref}}$ is the zero doppler, delayed versions of the reference signal:

$$\mathbf{S_{ref}} = \begin{bmatrix} \mathbf{s_{ref}} & \mathbf{Ds_{ref}} & \mathbf{D^2s_{ref}} & \cdots & \mathbf{D^{K-1}s_{ref}} \end{bmatrix} \tag{3.6}$$

with **D** a 0/1 permutation matrix that applies a single sample delay defined as

$$\mathbf{D} = \{d_{ij}\}_{i,j=1,\dots,N+R-1}, \qquad d_{ij} = \begin{cases} 1, & \text{if } i = j + 1 \\ 0, & \text{otherwise} \end{cases} \tag{3.7}$$

The columns of matrix **X** in 3.2 defines a basis for an M-dimensional interference subspace, where $M = (2P + 1)K$. Solving 3.2 yields

$$\alpha = (\mathbf{X^H X})^{-1}\mathbf{X^H s_{surv}} \tag{3.8}$$

After cancellation the surveillance signal becomes

$$\mathbf{S_{ECA}} = \mathbf{s_{surv}} - \mathbf{X}\alpha = [\mathbf{I_N} - \mathbf{X}(\mathbf{X^H X})^{-1}\mathbf{X^H}]\mathbf{s_{surv}} = \mathbf{Ps_{surv}} \tag{3.9}$$

where **P** is the projection matrix that projects the received vector $\mathbf{s_{surv}}$ in the subspace orthogonal to the interference space.

The major limitation of this algorithm is its high computational cost, $O[NM^2+M^2logM]$. This limitation is addressed by further modifications of the algorithm and is called ECA batches (ECA-B). Details of ECA-B can be found in [39].

3.3.2 Conjugate Gradient Least Squares algorithm

The Conjugate Gradient Least Squares (CGLS) algorithm has been identified as an effective filtering technique for signal processing in various fields. It is a least squares based refinement algorithm with the ability to vary the amount of cancellation performed through the limiting of the number of refinement iterations. This enables the ability to optimise the execution time by performing only the required amount of cancellation. The number of iterations to achieve sufficient self-interference cancellation can be estimated through trial and error because for a stationary transmitter and receiver setup the self-interference and clutter are reasonably constant over time. The CGLS algorithm [38, 40] is described below, it too involves solving the following minimisation problem:

$$\min_{\alpha} \left\{ ||\mathbf{s_{surv}} - \mathbf{X}\alpha||^2 \right\} \tag{3.10}$$

Let α_k be the parameter vector after the k-th iteration. The residual vector is $\mathbf{r}_k = \mathbf{s_{surv}} - \mathbf{X}\alpha_\mathbf{k}$, the negative gradient is $\mathbf{p}_k = \mathbf{X^T r}_k$ and $\gamma = ||\mathbf{Xp}_k||^2$.

Given an initial guess α_0, set: $\mathbf{r}_0 = S_{surv} - \mathbf{X}\alpha_0$
$\mathbf{p}_0 = \mathbf{X^T r_0}$
$\gamma_0 = ||\mathbf{p}_0||^2$
$\mathbf{s} = \mathbf{p}$
for $k = 1$ to m {
$\mathbf{q} = \mathbf{Xp}$
$\mathbf{a} = \gamma / ||\mathbf{q}||^2$
$\alpha = \alpha + \mathbf{ap}$
$\mathbf{r} = \mathbf{r} - \mathbf{aq}$
$\mathbf{s} = \mathbf{X^T r}$
$\gamma_{k-1} = \gamma_k$
$\gamma_k = ||\mathbf{s}||^2$
$\beta = \gamma_k / \gamma_{k-1}$
$\mathbf{p} = \mathbf{s} + \beta\mathbf{p}$
}

3.4 Noise and distortion theory

3.4.1 Non-linear distortion

In practice, all realistic components have some losses and become non-linear at very low and high signal levels. These non-linear effects set an effective minimum and maximum realistic power range, better known as the dynamic range, over which a component or network will operate as desired. While non-linear characteristics are desirable for functions such as amplification, detection and frequency conversion, they also lead to undesirable effects in RF circuits such as harmonic generation, saturation and intermodulation distortion [41]. These effects can lead to increased losses, signal distortion and interference.

Figure 3.3 shows a general non-linear network representation with an input and output voltage of v_i and v_o respectively. In general, the output response of a non-linear circuit can be modelled as a Volterra series which has a memory effect. However, for time independent systems, a non-linear system can be sufficiently modelled by a Taylor series expansion in terms of the input signal voltage:

$$v_o = a_0 + a_1 v_i + a_2 v_i^2 + a_3 v_i^3 + \cdots , \tag{3.11}$$

where the Taylor coefficients are defined as

$$a_0 = v_0(0) \qquad (DC) \tag{3.12}$$

$$a_1 = \left.\frac{dv_o}{dv_i}\right|_{v_i=0} \qquad (linear) \tag{3.13}$$

$$a_2 = \left.\frac{d^2v_o}{dv_i^2}\right|_{v_i=0} \qquad (squared) \tag{3.14}$$

$$\vdots$$

and further higher order terms. Depending on the non-linear network, particular terms of the expansion may have a dominance resulting in various different functions being obtained. The most common cases of importance are discussed below.

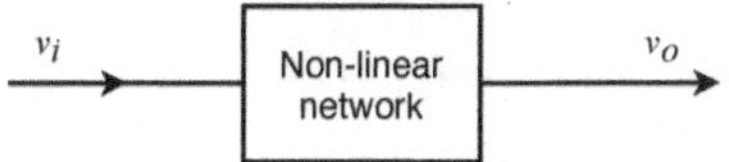

Fig. 3.3. A general non-linear network representation.

Gain compression

Linear amplifiers have a relatively fixed gain across a specified frequency range. A plot of the output power vs the input power presents a linear relationship (straight line graph), where the slope of the graph is the gain of the amplifier. As the input power continues to increase, there will be a point at which the amplifier goes into compression where any further increase in input power will not result in increasing output power. Instead the gain flattens and the amplifier is said to have reached saturation. The output response thus becomes non-linear and results in signal distortion, harmonics and intermodulation products. Physically this is a result of the instantaneous output voltage of an amplifier being limited by the bias voltage of the device.

A typical response (solid line) for an amplifier is shown in Figure 3.4. The straight line with unity slope/gain (dashed line) represents an ideal linear amplifier response. The linear operating range of the amplifier is quantified by the 1 dB compression point (P_{1dB}). The 1 dB compression point is defined as the input power level for which the output power level has decreased by 1 dB from the ideal linear response. This can be specified in terms of the input power level (IP_{1dB}) or the output power level (OP_{1dB}), typically chosen as the larger of the two [41, 42].

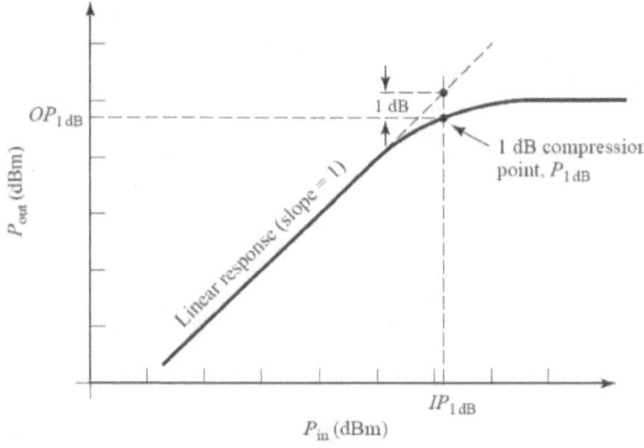

Fig. 3.4. Illustration of the 1 dB compression point (P_{1dB}) on a plot of input power vs output power [41].

Harmonic and intermodulation distortion

Harmonic distortion is a single tone distortion product that results from device non-linearity. For a single input frequency, ω_0, in a non-linear device, the output signal will consist of spurious harmonics of the input frequency of the form $n\omega_0$ where $n = 0, 1, 2, 3, \cdots$. The order of the harmonics are given by the frequency multiplier, i.e. $2\omega_0$ is the second order harmonic, $3\omega_0$ is the third order harmonic, etc.

Intermodulation distortion is a multi-tone distortion product that results from two or more signals are provided at the input to a non-linear device. The spurious intermodulation products generated by a non-linear device is mathematically related to the original input signals. The analysis of multiple tones can become very complicated; to demonstrate the principle, a two-tone analysis is used [41,42]. The output spectrum will consist of harmonics of the form

$$m\omega_1 + n\omega_2 \tag{3.15}$$

where $m, n = 0, \pm1, \pm2, \pm3, \cdots$. The various combinations of the two input frequencies is what are known as the intermodulation products. The order of a given intermodulation product is defined as $|m| + |n|$.

For the two-tone case, the resulting second order intermodulation products are $2\omega_1$, $2\omega_2$, $\omega_1+\omega_2$ and $\omega_2-\omega_1$. The third order intermodulation products are $3\omega_1$, $3\omega_2$, $|2\omega_1 \pm \omega 2|$ and $|2\omega_2 \pm \omega_1|$. This spectrum is shown in Figure 3.5. Intermodulation products can be removed through filtering, but of particular concern are the two third order products of $|2\omega_1 - \omega 2|$ and $|2\omega_2 - \omega_1|$, because these occur adjacent to the input tones and are usually too close to the fundamental frequencies be filtered out easily. The complete mathematical derivation of the intermodulation products can be found in [41].

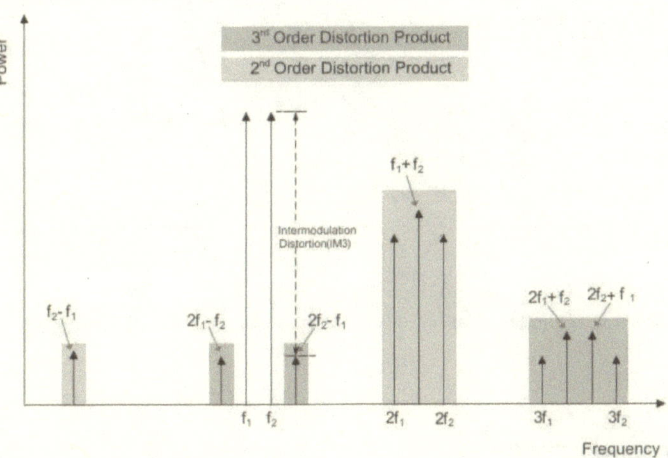

Fig. 3.5. Illustration of frequency spectrum with harmonics and second and third order intermodulation products [42].

Third order intercept point (IP_3)

Third order intermodulation products have a cubic relationship between the input and output voltages, i.e. there is an exponential relationship between the input and output voltages. Thus for small input powers the third order intermodulation products are small, but increase rapidly as the input power increases. Figure 3.6 shows a plot of the output vs input power for both first and third order intermodulation products.

Both first and third order responses will reach compression at high input powers. If the two idealised linear responses are hypothetically extended, they will intersect (dashed lines) as they have different slopes. This intersection point where the first and third order powers are equal is known as the third order intercept point and is denoted as IP_3. As with the 1 dB compression point, it can be specified as either an input power (IIP_3) or output power (OIP_3) and is typically chosen as the larger of the two. The IP_3 generally occurs at a higher power level than the 1 dB compression point, in practical devices typically 10 dB above the P_{1dB}. The IP_3 value is never achieved in practice as the device will saturate before this condition occurs, but is a useful indicator of the linearity of a device [41, 42].

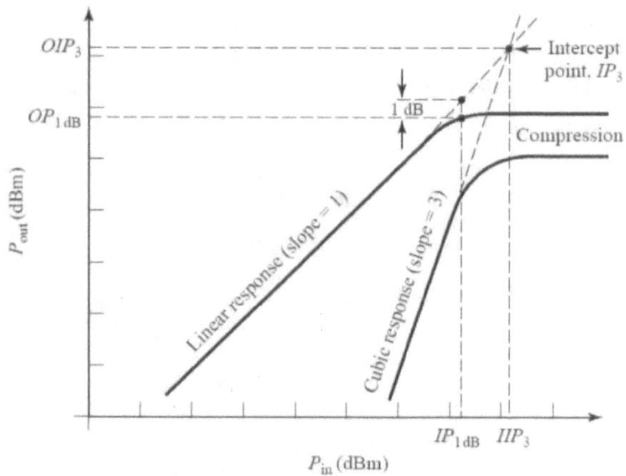

Fig. 3.6. Illustration of the third order intercept point (IP_3) on a plot of input power vs output power [41].

3.4.2 Dynamic range

The dynamic range of a system or component is defined as the operating range for which that system or component has desirable characteristics. For devices such as amplifiers this may be the power range limited by noise at the low end and by the compression point at the high end, in which case it is essentially the linear operating range of the system and is called the linear dynamic range (LDR). For devices such as low noise amplifiers or mixers the power range may be limited by noise at the low end and the maximum level of intermodulation distortion, in which case it is essentially the operating range with minimal acceptable spurious responses and is called the spurious-free dynamic range (SFDR). The LDR is defined as the ratio of P_{1db} to the noise level of the device or system, N_o. SFDR is defined as the ratio of the fundamental signal of interest to the strongest spurious signal in the output [41, 42].

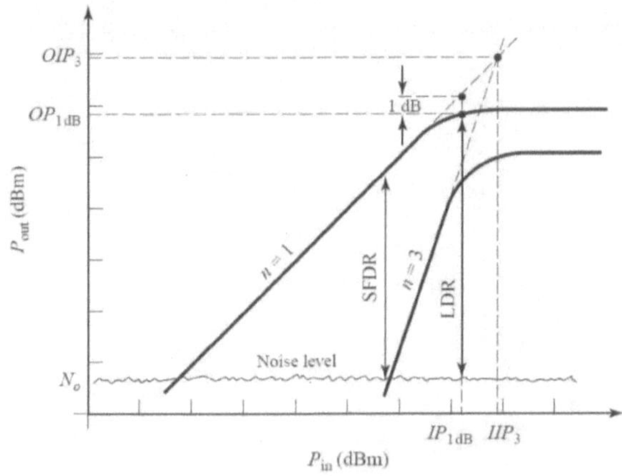

Fig. 3.7. Illustration of linear dynamic range (LDR) and spurious-free dynamic range (SFDR) on a plot of input power vs output power [41].

3.4.3 Mixers

Mixers are either active or passive 3 port non-linear devices which convert a signal from one frequency to another. An input RF signal is mixed with a local oscillator (LO) signal to produce an intermediate frequency (IF) output signal consisting of the sum $(f_{RF} + f_{LO})$ and difference $(f_{RF} - f_{LO})$ frequencies. The desired frequency of the two is selected using a filter. When the sum frequency is used the mixer is called an up-converter and when the difference frequency is used the mixer is called a down-converter. Since mixers are non-linear devices, the output will not only contain the sum and difference frequencies, but also harmonics and intermodulation products.

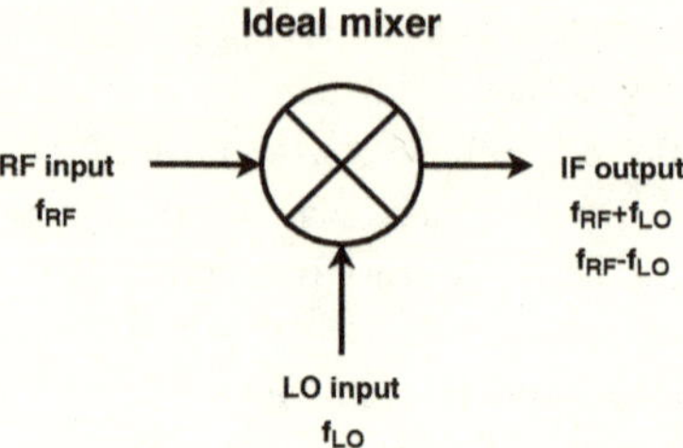

Fig. 3.8. An ideal mixer showing the frequency components of the inputs and output.

Chapter 4

STAR Demonstrator Design

This chapter details the methodology and design of a STAR demonstrator with narrowband self-interference cancellation. First an overview of the system is given, followed by a detailed discussion of the major components of the system.

4.1 System overview

This section provides a high level overview of the demonstrator and discusses the design approach and the major design choices which had to be made.

4.1.1 Design approach

It has become common practice to make use of multiple techniques for self-interference cancellation to enable STAR. The reason for this is that any single technique does not provide sufficient self-interference suppression to enable STAR, which is key to making such a system work. Therefore, the approach chosen for the demonstrator is to use a multi-layer cancellation scheme, in particular, a three layer system comprising passive suppression, analogue cancellation and digital cancellation.

Passive suppression is all about achieving as much isolation as possible between the co-located transmit and receive antennas. More isolation means less of the transmit signal leaks through into the receive antenna and hence less self-interference.

Analogue cancellation aims to remove as much self-interference as possible in the RF domain before the signal gets processed in the receive chain. This is to avoid non-linear signal distortion due to saturation of the receiver and to avoid damaging sensitive components in the receive chain.

Digital cancellation aims to suppress any residual interference remaining after RF cancellation by using techniques in the digital domain.

4.1.2 Operational frequency band

The Industrial, Scientific and Medical (ISM) radio bands are portions of the radio spectrum reserved internationally for the use of RF energy for industrial, scientific and medical purposes. In recent years the ISM bands have been dominated by wireless communication such as Bluetooth and WiFi.

The demonstrator will be required to transmit as well as receive signals. Therefore, an unlicensed ISM band has been chosen as the operational frequency band as it is freely available for use. It was decided that the South African 5.8 GHz band (5.725 GHz - 5.85 GHz) will be used. This frequency band allows a maximum power of 1 W peak equivalent isotropic radiated power (EIRP) with any modulation scheme and no further restrictions [43].

This band is chosen over the typical 2.4 GHz ISM band because there is lower usage (and hence, less external interference) than would typically be experienced due to Wi-Fi, Bluetooth and other devices/appliances which operate in the 2.4 GHz band.

4.2 Passive suppression

4.2.1 Passive suppression overview

The passive suppression layer can involve using a shared antenna or dual antenna setup. A shared antenna setup uses a single antenna and requires a circulator to isolate and direct the transmit and receive signals into their respective channels, while a dual antenna setup requires the use of two physically separated antennas for transmit and receive.

The transmit-receive isolation when using a shared antenna setup is dependent on the circulator performance and antenna matching, as these components contribute to the self-interference in this setup. In a dual antenna setup with a separate antenna for transmission and reception, the direct coupling between the antennas contributes to the self-interference and the isolation is dependent on frequency and physical separation and placement of the antennas.

The major drawback of a dual antenna system is the high amount of self-interference that is experienced when the transmit and receive antennas are co-located, compared to when the antennas are far apart with high path loss between them. However, A dual antenna setup offers the advantage of flexibility to use additional techniques to improve the transmit-receive isolation at the cost of additional hardware. For this reason, a dual antenna setup is chosen over a shared antenna setup.

The transmit and receive antennas of the proposed system will be co-located. When the antennas are co-located, physical separation cannot be well exploited, thereby limiting the amount of isolation obtained from path loss. Shielding and polarisation diversity are techniques which form part of high isolation antenna design, which is not within the scope of this project. Therefore, these techniques are not considered. Instead, an antenna array is utilised to create a directional transmitter, where the receiver is co-located in a null of the transmitter, resulting in increased isolation.

The antennas used are commercial Siretta Delta 7A Dual-band WiFi dipole antennas. These antennas are omnidirectional and have poor isolation, but have been chosen because of their low cost and availability. The antennas are operational at the 2.4 GHz and 5.8GHz ISM frequency bands. The detailed specifications of this antenna can be found in Table 4.1.

Fig. 4.1. The Siretta Delta 7A dual band WiFi/ISM antenna.

Table 4.1. Summary of antenna specifications [44]

Impedance:	50Ω
Operating frequencies:	2.4, 5.8 GHz
Gain:	1.5 dBi (2.4-2.5 GHz), 2.1 dBi (5.8 GHz)
VSWR:	<2.0:1
Polarisation:	Linear
Radiating element:	$\frac{1}{4}$ wave omni-directional dipole
Connector:	SMA male

Ideally, a high isolation transmit-receive antenna pair would be designed, but it is not the main focus of this project. The aforementioned WiFi antennas are sufficient for demonstrative purposes, as passive suppression techniques can be applied to improve the isolation between the transmit and receive antennas.

The simplest form of a dual antenna setup comprises a single transmit and single receive antenna, spaced some distance apart. This antenna configuration is used to obtain a baseline measure of performance. The transmit antenna is then replaced by an antenna array with which the receive antenna can be placed within a null of the transmit array.

In order to gain insight on the electromagnetic (EM) performance of the antenna configurations, simulation are conducted using Altair HyperWorks FEKO [45]. These two antenna configurations are discussed in more detail below along with their EM simulation results.

4.2.2 Single transmit and single receive antenna configuration

A single dipole has an omnidirectional radiation pattern in the azimuth/horizontal plane. If an identical receive antenna is co-located to such a transmit antenna in a system, there will be mutual coupling between the two antennas which is identified as self-interference. As previously discussed, in a dual antenna setup there is an inherent path loss which reduces the strength of self-interference. If the two antennas are placed in the far field of each other, Equation 3.1 can be used to estimate the amount of path loss. However, if the antennas are placed in the near field of each other, then the path loss is not predictable because the exact EM effects are unknown.

For this design, the antennas are chosen to be arbitrarily co-located with a physical separation of one and a half wavelengths which places the antennas in the near field. This physical separation is representative of a realistic case of co-located transmit and receive antennas and is to remain as a fixed parameter in the system.

The performance of the antenna pair is obtained from FEKO simulations. The far field radiation pattern of the transmit antenna as well as the S-parameters of the two antennas are simulated. The S-parameters gives the coupling between the two antennas (S_{21}) from which the transmit-receive antenna isolation is determined.

The simulations are setup in CADFEKO by constructing two wire dipoles with operating frequency of 5.8 GHz, separated by one and a half wavelengths, with a feed on the transmit antenna. Ideal wire dipoles are used because the exact physical characteristics of the real antennas are not known other than the information provided on the data sheet [44]. Thus the dipole antenna model is ideal and the actual measured results will differ. Therefore, all the simulations in this chapter serve as a theoretical demonstration of performance and should not be mistaken for an identical simulation of the actual hardware.

The simulations are configured to compute a 3D radiation pattern as well as the S-parameters over a frequency range of 200 MHz centred at 5.8 GHz. S_{21} is the simulated S-parameter of interest as it is used to compute the isolation. The simulation results are shown in Figure 4.2.

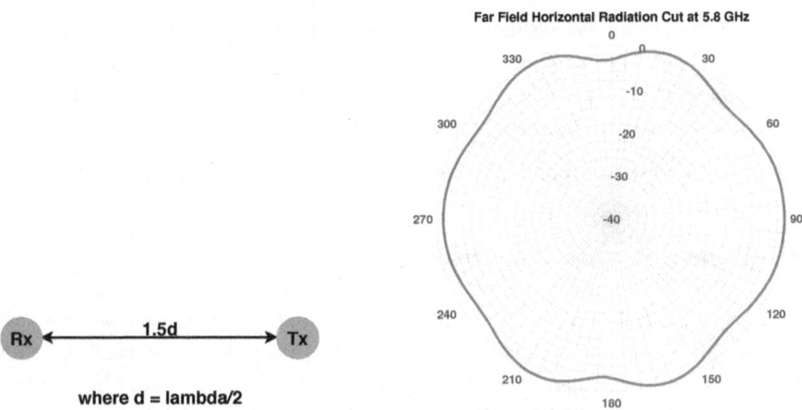

(a) The single transmit antenna configuration. (b) Horizontal cut of the transmit antenna radiation pattern (with receiver in place).

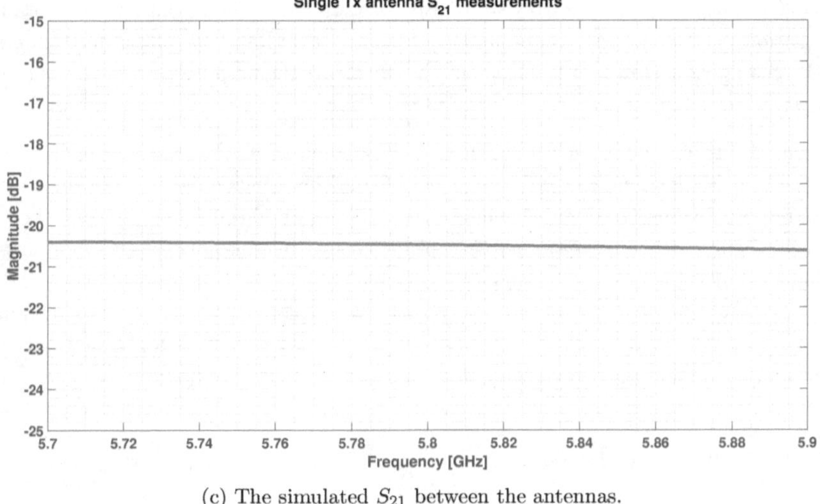

(c) The simulated S_{21} between the antennas.

Fig. 4.2. (a) The dual antenna configuration with single transmit antenna, (b) the FEKO simulated transmit antenna radiation pattern and (c) the simulated S_{21} for the configuration.

Figure 4.2b shows a horizontal cut of the far field radiation pattern. It is observed that the radiation pattern differs from an ideal dipole on its own (i.e. without the receive antenna placed nearby). This is due to the coupling between the two antennas.

Figure 4.2c shows the simulated S_{21} of the setup over a 200 MHz bandwidth. The S_{21} parameter indicates how much power is coupled from the transmit antenna to the receive antenna. The transmit-receive isolation is considered to be the amount of power loss from transmit to receive (which is also the direct coupling or self interference between the two antennas). Taking the absolute value of the S_{21} parameter gives an indication of the transmit-receive isolation. The simulated transmit-receive isolation is approximately 20.5 dB.

The actual radiation pattern and isolation is measured and discussed in Chapter 5.

4.2.3 Transmit antenna array

Depending on the application, the required transmit antenna radiation pattern would differ. For example, in a communication system one might require a narrow main beam for line of sight links and for a surveillance radar one might require a broader main beam. The required radiation pattern for a particular application would form part of a high isolation antenna design. For the purposes of this demonstrator, no radiation pattern requirements are imposed.

Producing a null in an antenna array's radiation pattern can be done using various techniques. The radiation pattern depends on a number of factors including: the actual antenna elements used, the geometric location of the elements in the array, the amplitude, and the phase of the feed signal to each antenna.

In Section 3.1.3 techniques such as the Schelkunoff polynomial method are discussed. Most such techniques which allow one to synthesise a desired null in a particular position in a radiation pattern, require feeding the array elements with different amplitudes and phases. The number of elements being used affects the number of degrees of freedom one has to manipulate the radiation pattern. Theoretically, one can compute what these required amplitude and phase weights are, but practically they can be complex to implement as additional hardware and control is required. However, there are also ways to produce nulls using arrays without the need for amplitude and phase weighted elements.

Uniform Linear Array (ULA)

A uniform linear array is the most basic form of an array which has been well studied and widely documented. It is constructed using two or more antenna elements all placed on a single axis. ULAs are not only simple to construct, but they are versatile as well. The antenna elements, the number of array elements, the spacing between the elements and the amplitude and phase of the feed signals can all be manipulated to tailor a desired radiation pattern.

Of particular interest is the production of nulls in the radiation pattern, such that the receive antenna can be placed therein to maximise the transmit-receive isolation. Pattern multiplication is used to obtain the far field radiation pattern from the array factor (AF) and element pattern. However, the far field radiation pattern does not represent the near field radiation. In order to produce nulls in the near field at a particular location, one must look to the simplest form of a signal being transmitted.

RF signals are typically sinusoidal with some amplitude and phase. Using the principle of destructive interference of signals, if two equal amplitude sinusoidal wave forms that are 180°out of phase are combined, they will cancel each other out. Each element in the array transmits a signal and the superposition of these signals is what produces the resulting far field radiation pattern.

In order to ensure that there is a near field null at the receiver, the feed signals to each element, the array element positions as well as the receiver position is important. The spacing should be such that cancellation of the transmit signals occur at the receiver position. e.g. If there are two elements, the elements should be a half wavelength (180°phase difference) apart from each other, with the receiver on the same axis of the array. In addition to this, each of the array elements needs to be coherently fed with the same amplitude and phase signal to ensure destructive interference at the receiver location. This is illustrated in Figure 4.3. If the elements are fed with different phases, the spacing between the elements would have to change in order to get destructive interference.

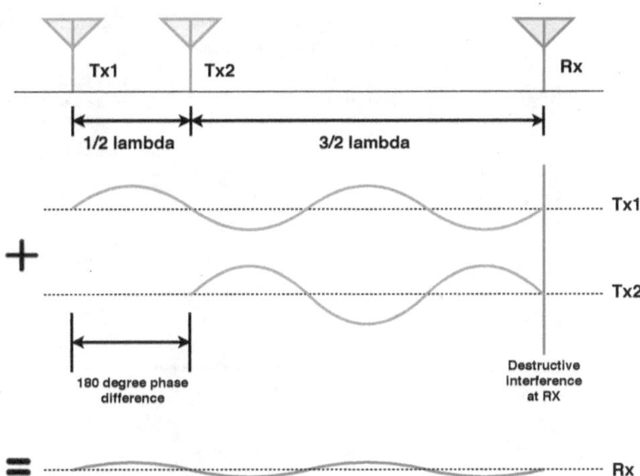

Fig. 4.3. Illustration of near field cancellation principle using a linear antenna array with half a wavelength spacing between the elements. The receiver is on the same axis as the linear array.

The effects of increasing the number of elements include making the array factor more directional, increasing gain and an increasing number of side lobes. Three cases, using two, three and four element linear arrays are simulated in FEKO. The results are shown in Figure 4.4.

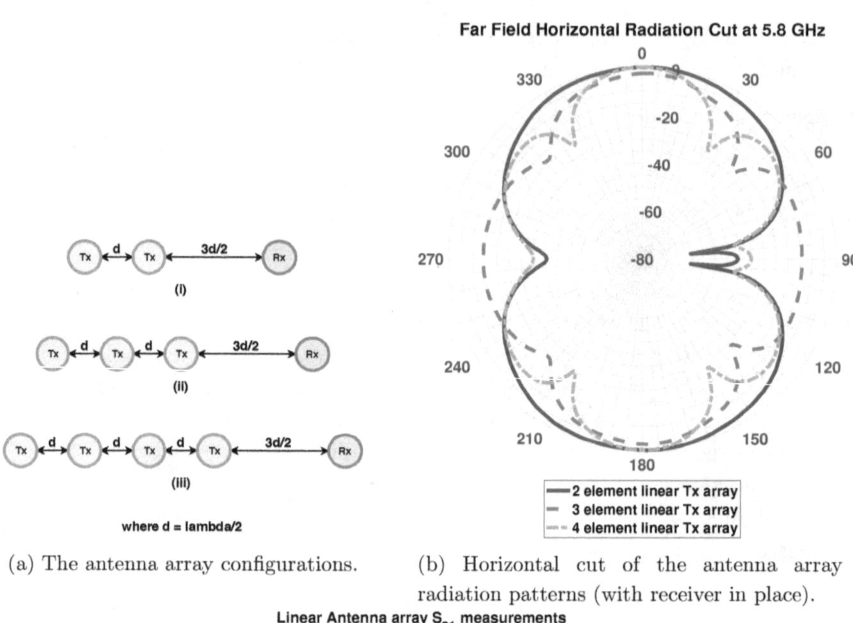

(a) The antenna array configurations.

(b) Horizontal cut of the antenna array radiation patterns (with receiver in place).

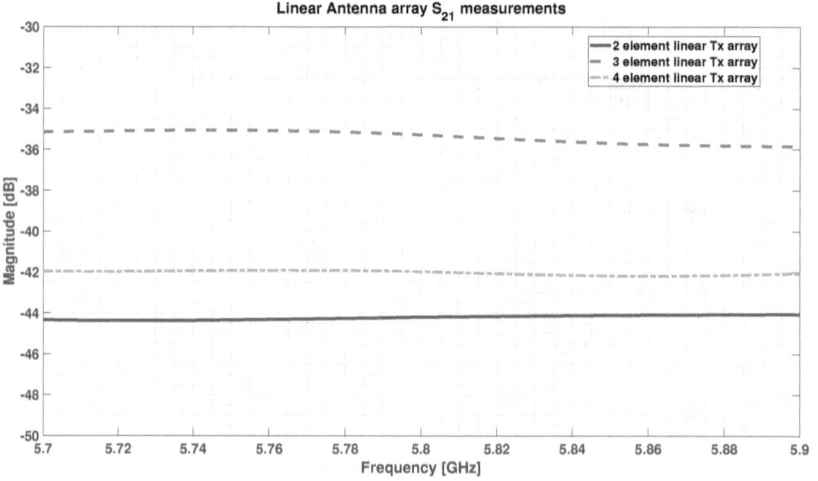

(c) The simulated S_{21} for the two, three and four element array configurations.

Fig. 4.4. (a) The dual antenna configurations with transmit arrays, (b) the FEKO simulated transmit antenna array radiation patterns and (c) the simulated S_{21} for each of the array configurations.

Figure 4.4a illustrates the antenna configuration used for each of the simulations. Each of the transmit antennas are fed coherently with identical amplitude and phase. The far field radiation patterns were computed and are shown in Figure 4.4b. It is observed that there is a big difference in the magnitude of the nulls at $\pm 90°$ when using an odd and even number of elements in the transmit array. This is because when an odd number of elements is used with half a wavelength spacing between the elements, there is a residual signal component that doesn't get cancelled out. Thus an odd number of elements is not suitable when using a half wavelength spacing. This is further emphasised by the coupling between the transmit array and the receive antenna. The transmit-receive coupling is indicated by the S_{21} parameter shown in Figure 4.4c and is a measure of the transmit-receive isolation. Using two or four elements gives 44 dB and 42 dB of transmit-receive isolation respectively while the three element array gives a much lower isolation of 35 dB.

Alternative configurations such as circular arrays with progressive phasing or asymmetric arrays could also be used. However, a linear array, in theory, can provide more than 20 dB more isolation compared to using a single omnidirectional transmit antenna and is sufficient for demonstrative purposes. Since sufficient antennas and a suitable power splitter are available, a four element linear array is used for the demonstrator.

4.3 Analogue cancellation

4.3.1 RF canceller overview

Analogue cancellation can be done in various ways, each with their own benefits and drawbacks. For example, using a balun and/or delay lines have the benefit of providing cancellation over a wider bandwidth compared to a phase shifter, which can only provide narrowband cancellation. Since wideband cancellation is not required, a digital attenuator and phase shifter are used, as these are sufficient to demonstrate the concept of narrowband analogue cancellation.

The HMC1122LP4ME 6-bit digital step attenuator and HMC649ALP6E 6-bit digital phase shifter from Analog Devices were chosen. These components were available at low cost and in small form factor, surface mount packages. Due to the high cost of evaluation boards for these components and comparatively lower cost of prototyping printed circuit boards (PCBs), the latter option was chosen. The resolution of the digital attenuator and phase shifter is 0.5 dB and 5.25 degrees respectively.

The design of the analogue cancellation layer requires splitting a signal from the transmit chain and applying an attenuation and phase shift to this signal, then adding it to the received signal. the transmit signal is the ideal signal to use as a cancellation signal because the self-interference signal can be modelled as an amplitude scaled and delayed version of the transmit signal. The more accurate the cancellation signal is, the better the performance of the analogue cancellation will be. Therefore, the resolution of attenuation and phase adjustment are vital to performance.

Since the exact amplitude and phase of the received signal is not constant and will change based on various factors including the transmit signal power and multipath effects, the amount of attenuation and phase shift applied to generate the cancellation signal must be tunable. For best performance, the attenuation and phase shift should be automated and controlled via feedback, but automation is not within the scope of this project. Instead, these components will be controlled via a manually actuated control module.

A diagram of the analogue cancellation circuit is shown in Figure 4.5. The transmit signal goes into port 1 and is split by a two-way power divider, the first path into the transmit antenna (port 3) and the second path into the cancellation channel. The cancellation channel contains a digital step attenuator module as well as a digital phase shifter module which are controlled by a separate control module. After passing through the attenuator and phase shifter, the cancellation signal is combined with the signal from the receive antenna (port 4).

The control module is used to tune the cancellation signal so that it adds destructively with the self-interference in the received signal, before it moves on to the rest of the receive chain which is connected to port 2.

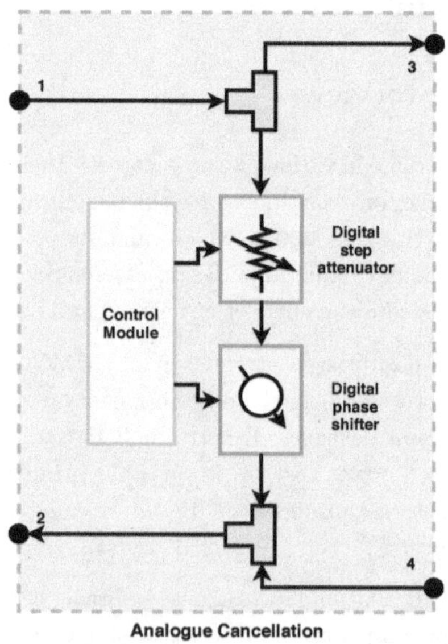

Fig. 4.5. A block diagram of the analogue cancellation circuit.

4.3.2 Digital attenuator module

The digital attenuator module is a prototype PCB module designed using the Analog Devices HMC1122LP4ME surface mount 6-bit digital step attenuator. The HMC1122LP4ME operates from 0.1 GHz to 6 GHz, has an attenuation control range of 31.5 dB and has a resolution of 0.5 dB. It can be controlled via a serial or parallel interface. The prototype design is based on an evaluation board presented in the device datasheet [46].

The PCB was designed for operation at 5.8 GHz using Altium Designer and rapidly prototyped. It was manufactured by TraX Interconnect [47] on FR4 substrate with a dielectric constant of 4.4 and a thickness of 1.5 mm. The bottom of the board as well as the unused parts of the board are filled with a copper ground plane. The top and bottom ground planes are connected using suitable via stitching with 1/20th wavelength spacing. The PCB is interfaced via 50 ohm SMA male connectors for the RF signals and a pin header for the digital signals. A Thru calibration is designed into the module at the bottom so that the attenuator itself can be characterised. The design is shown in Figure 4.6a and the manufactured prototype in Figure 4.6b.

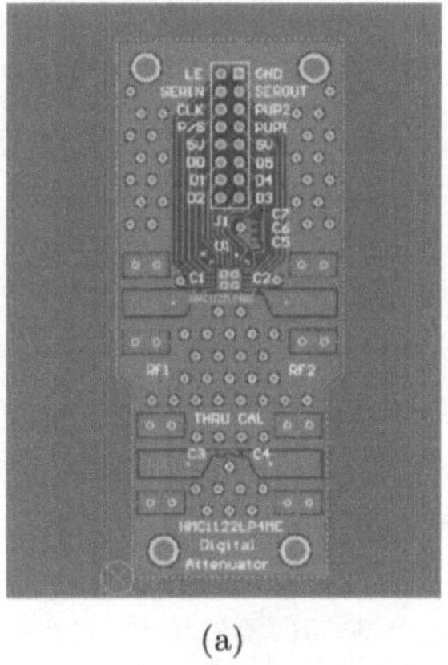

(a)

(b)

Fig. 4.6. (a) The digital attenuator module design and (b) the prototype digital attenuator module.

After manufacturing and assembly, the digital attenuator module was characterised by measuring the S-parameters using a vector network analyser (VNA). Using the Thru calibration, the RF transmission line responses are removed leaving just the response of the attenuator itself. These responses are shown in Figure 4.7.

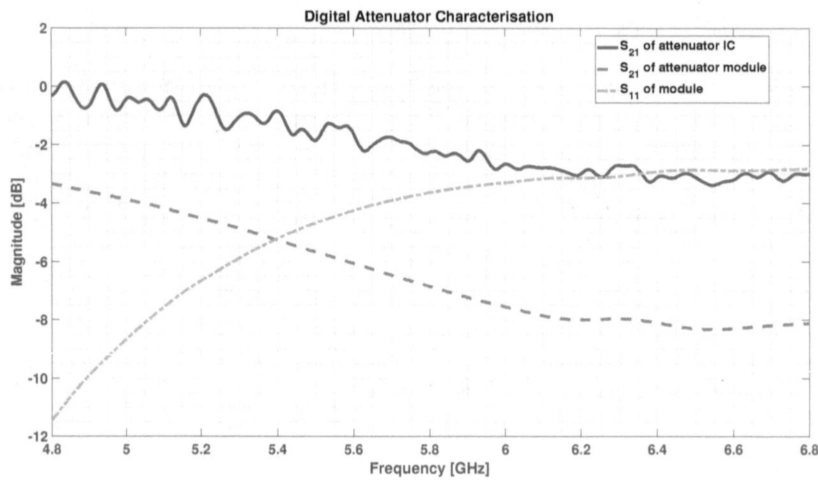

Fig. 4.7. Digital attenuator S-parameter characterisation.

From Figure 4.7, the return loss (S_{11}) of the module is approximately 4 dB and the insertion loss (S_{21}) is greater than 6 dB at 5.8 GHz. After calibrating out the transmission lines the insertion loss is approximately 2 dB. The typical rated return loss is 21 dB and the typical rated insertion loss of the attenuator is 2 dB at the required frequency. The non-ideal return loss and high insertion loss indicates a problem with the PCB. Since the insertion loss of the attenuator is correct, it means that the further losses are likely related to the transmission lines.

After detailed inspection and EM simulation using AWR Microwave Office [48], it was identified that the cause of the loss is impedance mismatch between the 50 ohm transmission line on FR4 and the component pad. The solution would thus be to redesign the PCB using a stepped impedance transformer to obtain the proper matching and reduce the overall insertion loss of the module.

4.3.3 Digital phase shifter module

The phase shifter module is a prototype PCB module designed using the Analog Devices HMC649ALP6E surface mount 6-bit digital phase shifter. The HMC649ALP6E operates from 3 GHz to 6 GHz, provides 360 degree phase coverage with a resolution of 5.625 degrees. It is controlled via a parallel interface. The prototype design is based on an evaluation board presented in the device data sheet [49].

Similarly to the digital attenuator, the PCB was designed for operation at 5.8 GHz using Altium Designer. It was manufactured by TraX Interconnect [47] on FR4 substrate with a dielectric constant of 4.4 and a thickness of 1.5 mm. The bottom of the board as well as the unused parts of the board are filled with a copper ground plane. The top and bottom ground planes are connected using suitable via stitching with 1/20th wavelength spacing. The PCB is interfaced via 50 ohm SMA male connectors for the RF signals and a pin header for the digital signals. A Thru calibration is designed into the module at the bottom so that the phase shifter itself can be characterised. The design is shown in Figure 4.8a and the manufactured prototype in Figure 4.8b

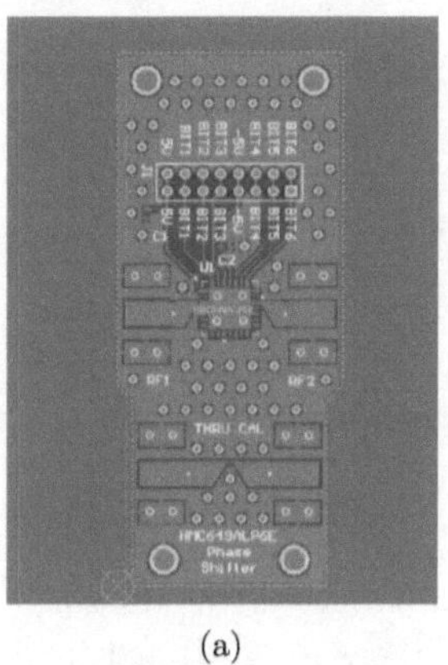

(a)

(b)

Fig. 4.8. (a) The digital phase shifter module design and (b) the prototype digital phase shifter module.

After manufacturing and assembly, the digital phase shifter module was characterised by measuring the S-parameters using a vector network analyser (VNA). Using the Thru calibration, the RF transmission line responses are removed leaving just the response of the phase shifter itself. These responses are shown in Figure 4.9.

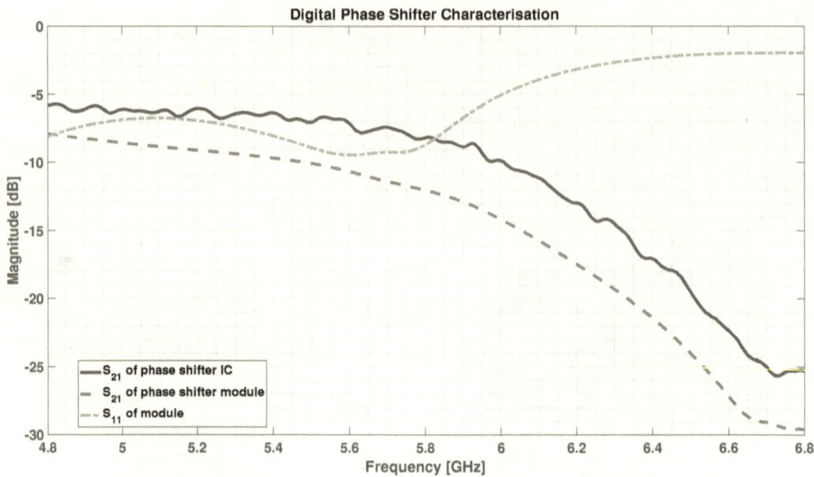

Fig. 4.9. Digital phase shifter S-parameter characterisation.

From Figure 4.9, the return loss (S_{11}) of the module is approximately 6 dB and the insertion loss (S_{21}) is greater than 14 dB at 5.8 GHz. After calibrating out the transmission lines the insertion loss is around 7 dB. The typical rated return loss is 13 dB and the typical rated insertion loss of the attenuator is 8 dB at the required frequency. As with the attenuator module, the return loss is not ideal and the insertion loss is quite high. The problem is also due to impedance mismatch and the solution is to use a stepped impedance transformer for proper impedance matching.

4.3.4 Parallel control module

The digital RF attenuator and phase shifter modules both require 6-bit binary control inputs. Since the system is experimental and not adaptive, manual control inputs are required. A simple control module circuit was designed on veroboard using switches to provide the binary inputs. Single pole double throw (SPDT) dual inline package (DIP) switches are used for compactness. Light emitting diodes (LEDs) are used as on/off indicators for each of the switches.

The control circuitry as well the attenuator and phase shifter modules are powered via 5 V dual supply voltage regulator circuit using the 7805 and 7905 series voltage regulators.

4.3.5 Complete RF canceller

The complete RF canceller requires the aforementioned modules to be connected together and to the control module. SMA T-splitters are used for the signal splitting and signal combining. For splitting a signal from the transmit chain, especially since the system will be using low power, the T-splitters are advantageous as a fairly strong cancellation signal is required to cancel out the self-interference. The complete RF canceller is shown in Figure 4.10.

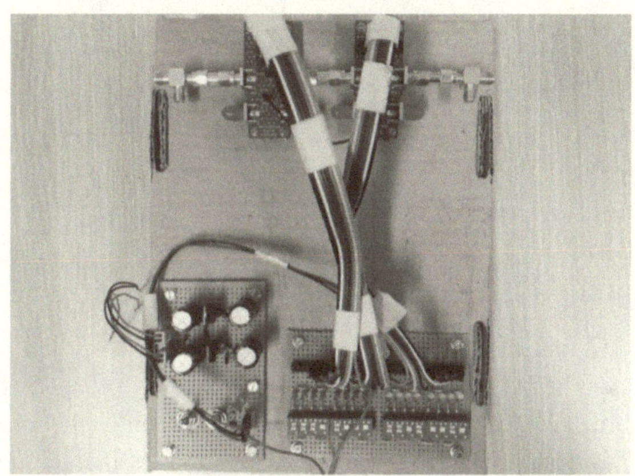

Fig. 4.10. Image of the complete analogue cancellation hardware.

The two modules have been individually characterised, but since they are to be connected together, the combined response is important. The combined response (excluding the SMA T-splitters) is shown in Figure 4.11.

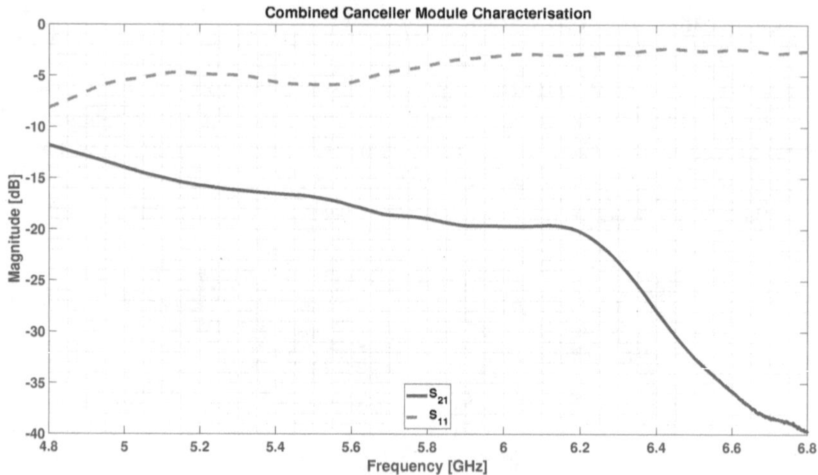

Fig. 4.11. Combined RF canceller S-parameter characterisation.

The combined return loss is approximately 4 dB and the combined insertion loss is approximately 18 dB when the digital attenuator is set to the minimum of 0 dB attenuation. Following on from the individual response, the return loss is not ideal and the insertion loss is quite high due to the impedance mismatches.

The canceller module was also tested on the VNA to see if it operates correctly, by observing the S_{21} and phase response as the control inputs are changed. Both the attenuator and the phase shifter performed according to specification. Since the canceller is operational even with the impedance mismatches, a second iteration design was not necessary. And with the system being experimental, the additional losses are not critical for the demonstrator.

4.3.6 Analogue cancellation simulations

To theoretically validate the viability of the RF canceller in cancelling out self-interference, a system level design was simulated in AWR Microwave Office. System blocks were used to model each of the blocks in Figure 4.5.

In order to simulate the analogue cancellation, the transmit and receive antennas or their isolation needs to be modelled as well. The antenna isolation (passive suppression) is modelled as a fixed attenuation of 30 dB. The digital attenuator and phase shifter are modelled with their measured insertion losses. The simulation block diagram is shown in Figure 4.12. test probes and conversion blocks are used where necessary to obtain measurements at various point in the system. The greyed out blocks are used for the time domain simulations in place of the network analyser block.

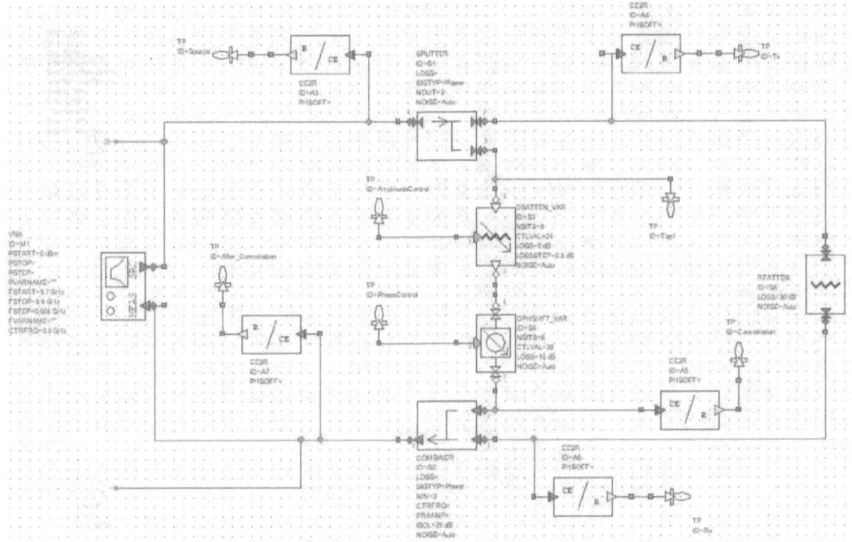

Fig. 4.12. AWR Microwave Office simulation system block diagram.

However, the attenuator and phase shifter both have specified typical and maximum operational error as well. These errors are taken into account for these simulations to gather insight of the best, typical and worst case performance. The rated errors are specified in Table 4.2.

Table 4.2. Component rated error specifications. [46,49]

Component:	Attenuator	Phase shifter
Typical error:	± 0.1 dB	-10 degrees
Maximum error:	-	+15/-32 degrees

The simulations are done in the time domain to visualise the cancellation and identify suitable control inputs for the digital attenuator and phase shifter. Thereafter, the S_{21} parameter is simulated. This is done for each of the best, typical and worst cases. The three scenarios are defined as:

- Best case: with no error
- Typical case: with typical errors
- Worst case: with maximum errors

The time domain simulations are shown in Figure 4.13 and the corresponding S-parameter simulations for each case in Figure 4.14.

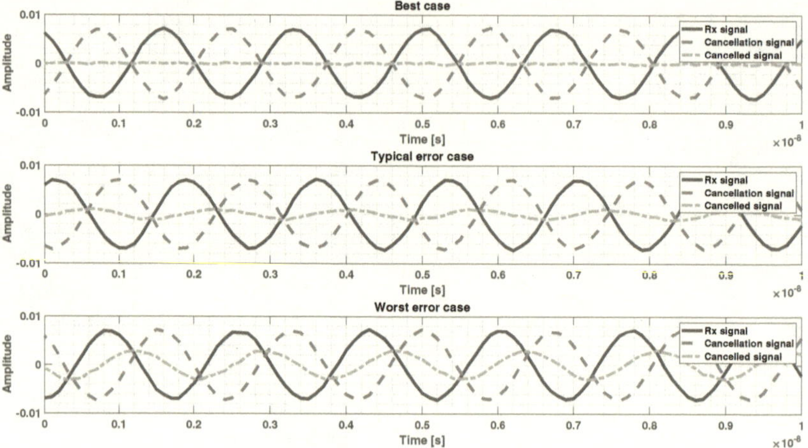

Fig. 4.13. Time domain analogue cancellation simulations.

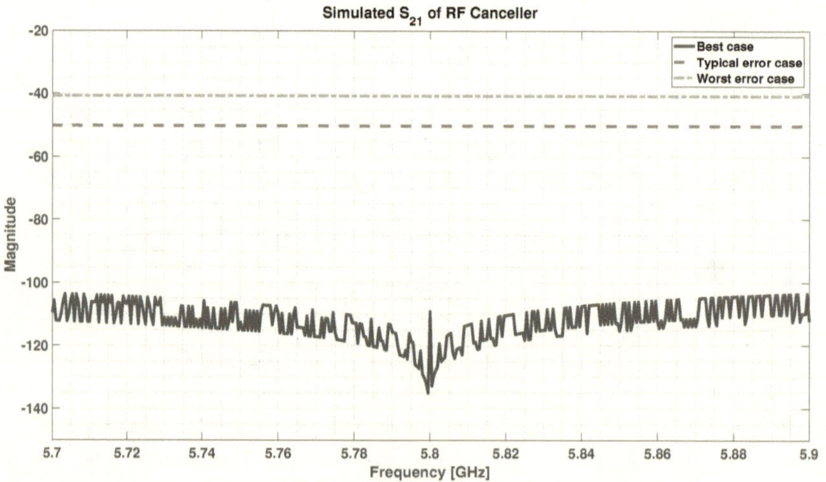

Fig. 4.14. S-parameter analogue cancellation simulations.

In Figure 4.13, the blue signal is the received signal representing the self-interference, the red signal is the cancellation signal that is manipulated with the attenuator and phase shifter, and the green signal is the signal after cancellation. In the best case scenario the self-interference signal is almost completely cancelled, in the typical case it is mostly cancelled and in the worst case it is only partially cancelled.

Looking at Figure 4.14, the S_{21} parameter gives a much clearer indication of the cancellation performance. Since there is 30 dB of passive suppression, any further cancellation is a result of the analogue cancellation hardware. The best case performance exceeds 100 dB, indicating near perfect cancellation. The typical case adds approximately 20 dB of cancellation, and the worst case scenario adds approximately 10 dB of cancellation. This shows that the accuracy of the cancellation signal plays a big role in the performance of analogue cancellation. The important observation from these simulations is that it proves theoretically that the analogue cancellation technique works.

4.4 Digital cancellation

4.4.1 Digital cancellation overview

The digital cancellation layer will comprise of adaptive filtering in the digital domain. Currently, Matlab implementations of adaptive filtering algorithms exist within the UCT Radar Remote Sensing Group (RRSG). These algorithms have been developed for passive radar projects within the research group. The methods used for these adaptive filters are the Conjugate Gradient Least Squares (CGLS) method and Extensive Cancellation Algorithm (ECA). These filters aim to suppress interference by removing the transmit reference signal from the received signal. These existing Matlab filter implementations will be adapted to work with the STAR demonstrator.

As is the case for analogue cancellation, optimisation is not a requirement for the digital cancellation. Therefore, the digital cancellation is not performed in real time, but rather on recorded signal data. However, should the digital cancellation be performed in real time, the CGLS algorithm would be more suitable. This is due to the ECA method having a high memory footprint and comparatively long computation time.

Since this is to be done in the digital domain, the analogue signals which feed into this layer of cancellation will need to be digitised. A Software Defined Radio (SDR) system will be considered for digitising and recording signals to a computer with Matlab. Digital cancellation can then be applied on the recorded signals.

4.4.2 Digital cancellation simulation

The digital cancellation should suppress any components of the coherently recorded reference signal that is present in the surveillance (receive) signal. Any signal that is not present in the reference should remain unaffected after digital cancellation. In order to verify that the digital cancellation implementations work, an experiment was conducted.

A 4 MHz chirp signal centred at 1.8 GHz was recorded coherently on both the reference and receive channels using the SDRs. Further details regarding the procedure of recording signal are discussed later on in Section 5.5.

This receive signal is regarded as the self-interference signal. An arbitrarily frequency shifted copy of this same signal was generated to simulate a signal of interest. These two signals were then combined to simulate a realistic received signal with both desired and undesired components. Digital cancellation was then performed on this combined signal. The expected result is that the interference signal should be suppressed and the received signal should not get suppressed.

The frequency spectrum of each of these signals are shown in Figure 4.15. After digital cancellation was done using the ECA algorithm, it is observed that the interference component has been suppressed while the signal of interest has not been suppressed. A similar result was obtained when using the CGLS algorithm.

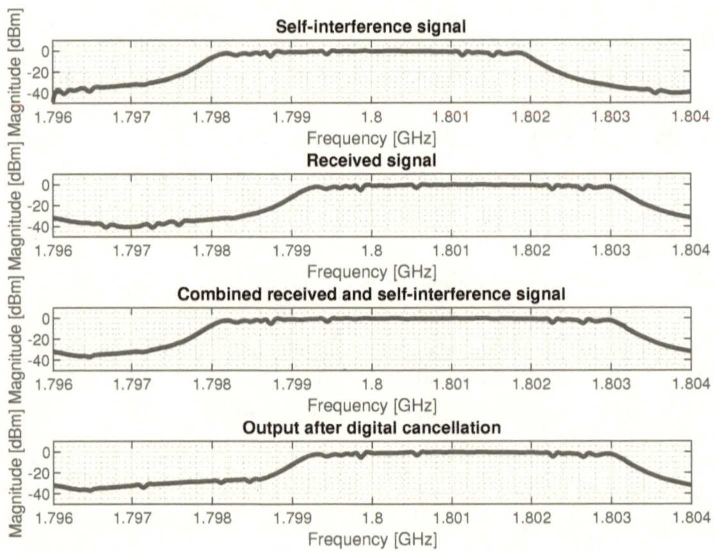

Fig. 4.15. Digital cancellation test showing the suppression of the self-interference component of the signal.

This result confirms that the digital cancellation is effective at suppressing self-interference.

Chapter 5

Results and Discussion

This chapter details the experiments carried out and presents the measured results obtained from experimentation. The measured results are analysed and discussed in detail in order to evaluate the effectiveness of the multi-layer self-interference cancellation STAR demonstrator.

5.1 Experimentation overview

5.1.1 Experimental environments

Various experiments were carried out to measure the performance of the STAR demonstrator and evaluate its practical viability. These experiments were carried out in two different environments: an anechoic chamber and a laboratory.

The anechoic chamber measurements characterise performance in an ideal environment with no multipath effects or external interference. The University of Cape Town (UCT) does not have an anechoic chamber facility. However, the Stellenbosch University does have a well equipped anechoic chamber and measurements were performed therein. The rectangular anechoic chamber is 9.1 m long, 5.5 m wide and 3.6 m in height. The interior of the chamber is shielded using an aluminium Faraday cage covered with absorber pyramids. A commercially integrated control system provides support for various measurement types including spherical near-field, planar near-field and conventional far-field measurements. Potentially supported frequencies range from 1 GHz to 26.5 GHz. Further information regarding the anechoic chamber can be found in [50]. An image of the chamber can be seen in Figure 5.1a.

Laboratory measurements are performed in the Microwave laboratory at UCT. The laboratory is not a controlled environment, has no RF shielding and is subjected to radiation including Bluetooth, cellular and WiFi radiation. The laboratory houses computers and measurement equipment and various other tools, making it quite a cluttered environment. Therefore, the laboratory measurements are representative of a typical real-world environment where multipath and external interference is likely to be present. This gives a more practical performance evaluation of the system. An image of the laboratory can be seen in Figure 5.1b.

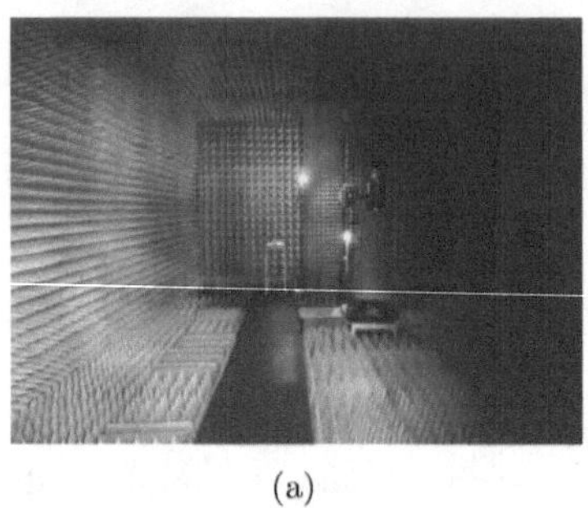

(a)

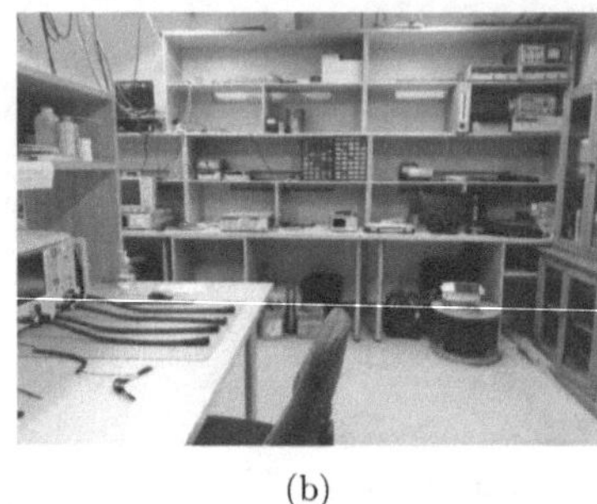

(b)

Fig. 5.1. (a) Image of the anechoic chamber and (b) Image of the laboratory environment.

5.1.2 Measurement equipment

The following equipment was used to procure the measured experiment results presented in this chapter:

- Agilent N5247A PNA-X Vector Network Analyser
- Agilent N9912A FieldFox RF Analyser
- Hewlett Packard 8350B/83592A Sweep oscillator and RF Plug-in
- Agilent E3620A Dual Output Power Supply
- Rohde & Schwarz SMF-100A Signal Generator

5.2 Antenna radiation pattern measurements

The far field radiation pattern of the dipole antenna and linear antenna array were measured in an anechoic chamber to verify the radiation patterns of the antennas. Results were obtained using a spherical near field scan and then transformed mathematically to the far field.

In theory, a single dipole antenna should have an omnidirectional radiation pattern, and the zero-phased linear dipole array should have broadside main beams facing forward and backward (0° and 180°) and nulls on either side (±90°) as shown in simulation. A horizontal and vertical cross sectional cut for each radiation pattern is shown along with the respective simulated radiation pattern in Figure 5.2.

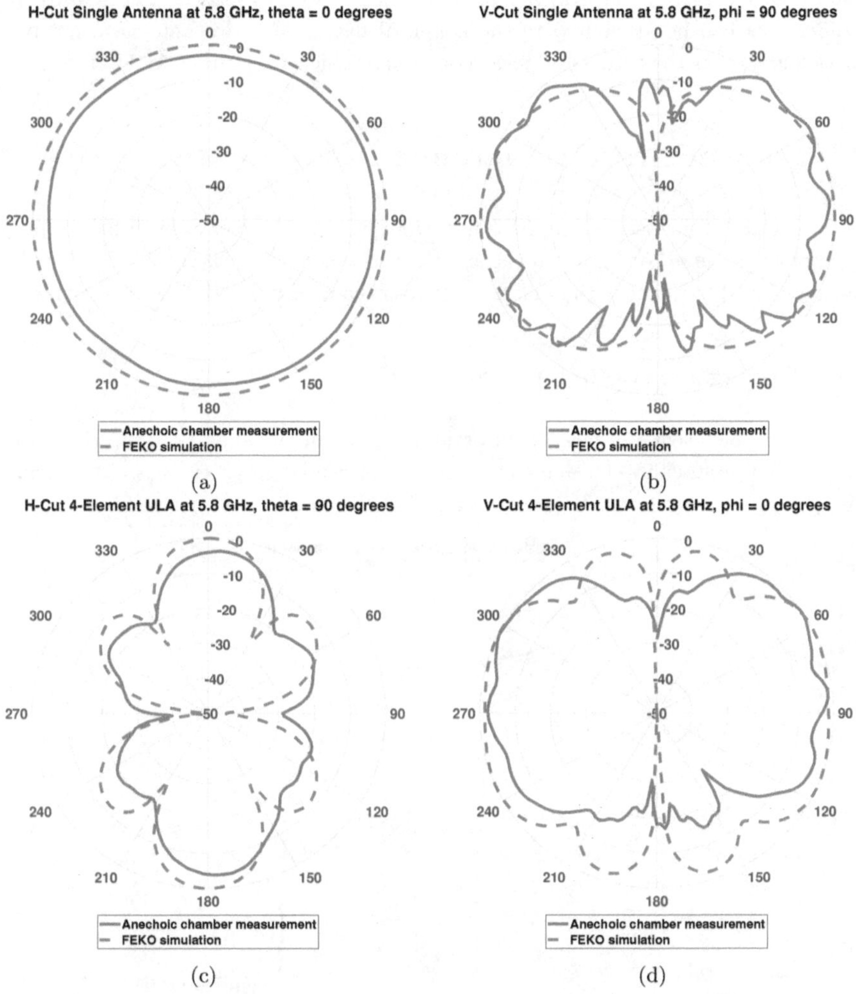

Fig. 5.2. Measured radiation pattern cuts in anechoic chamber: (a) Horizontal cut of single antenna radiation pattern, (b) Vertical cut of single antenna radiation pattern, (c) Horizontal cut of linear antenna array radiation pattern and (d) Vertical cut of linear antenna array radiation pattern

The measured radiation pattern of a single antenna closely resembles an omnidirectional pattern in the horizontal plane (Figure 5.2a). In the vertical plane (Figure 5.2b), it also closely resembles the simulated pattern. The measured antenna array radiation pattern also resembles the simulated patterns in both the horizontal (Figure 5.2c) and vertical (Figure 5.2d) planes. The differences can be attributed to the non-ideal nature of a real antenna compared to a simulation as well as the antennas being mass produced general purpose WiFi antennas.

5.3 Transmit-receive antenna isolation

The performance of the passive suppression layer is quantified by measuring S-parameters. The (S_{11}) parameter is used to verify that the antennas are resonant at the right frequency. The S_{21} parameter is used to quantify the isolation between the transmit and receive antennas.

5.3.1 Reflection coefficient (S_{11})

The reflection coefficient is a measure of performance of an antenna. The reflection coefficient parameter of the antenna and antenna array was measured to check that the antennas are radiating at the correct frequency which is 5.8 GHz. The result is shown in Figure 5.3.

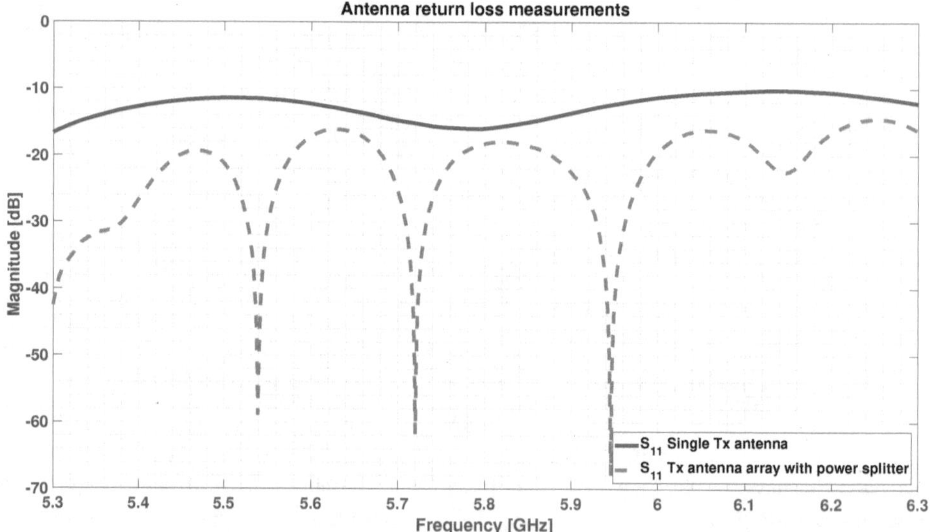

Fig. 5.3. Measured S_{11} of a single antenna and of the linear antenna array

From Figure 5.3, it can be seen that the S_{11} is well below -10 dB (which is the acceptable S_{11}

parameter value for a good antenna, i.e. the antenna radiates at least 90% of the input power) over a very wide band of at least 1 GHz, for both the single antenna as well as for the linear antenna array which uses a power splitter to feed each of the elements in the array. Thus the antennas are suitable for use at a centre frequency of 5.8 GHz.

5.3.2 Transmission coefficient (S_{21}) and Isolation

The isolation between two antennas can be defined as the difference in signal power between the transmit antenna's input and the receive antenna's output. The transmit-receive isolation was obtained from the transmission coefficient S_{21}, i.e. the amount of power measured at the receive antenna relative to the amount of power supplied to the transmit antenna. By inverting the sign of this measurement, the isolation between the two antennas can be obtained.

The isolation was measured both in an anechoic chamber as well as in a laboratory (multipath) environment using a single antenna and the antenna array. The measured S_{21} graphs over a 200 MHz bandwidth are shown in Figure 5.4 and the isolation summarised in Table 5.1.

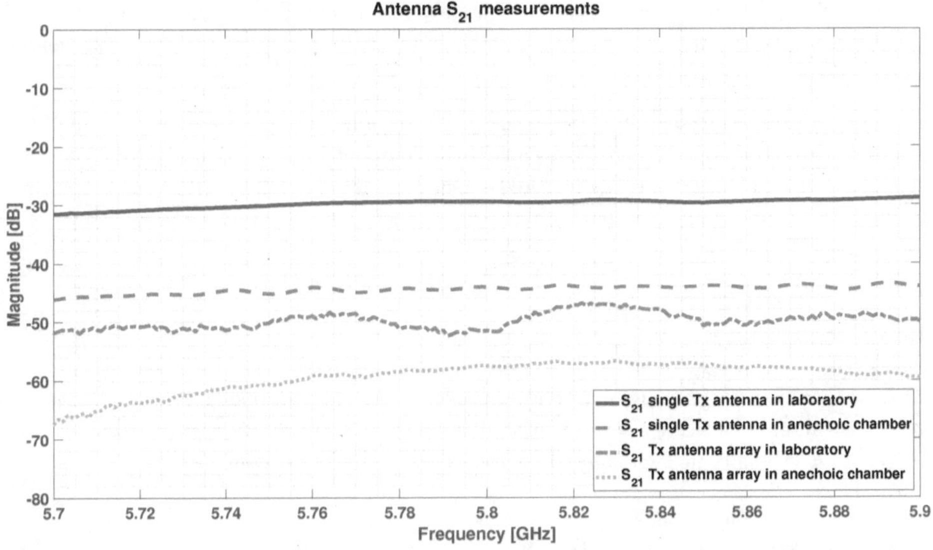

Fig. 5.4. Measured S_{21} of single antenna and linear antenna array in anechoic chamber and laboratory.

Table 5.1. Summary of isolation measurements

	Isolation [dB]	
	Anechoic chamber	**Laboratory**
Single antenna	44	29.4
Antenna array	57.5	51.5
Improvement	**13.5**	**22.1**

From Figure 5.4 it can be observed that there is significant differences between the anechoic chamber and laboratory measurements. The isolation between the transmit antenna and receive antenna is 44 dB in the anechoic chamber and 29.4 dB in the laboratory environment. Furthermore, the isolation between the transmit antenna array and receive antenna is 57.5 dB in the anechoic chamber and 51.5 dB in the laboratory environment. This amounts to the antenna array providing an additional 13.5 dB in the anechoic chamber and an additional 22.1 dB in the laboratory environment compared to using a single transmit antenna.

This agrees with the theory that the linear array configuration produces nulls on either side of the array along the axis on which the elements are placed. By placing the receive antenna in one of these nulls, the transmit-receive isolation will increase as shown by both simulation and measured results.

The large difference in isolation between the two environments can be attributed to multipath effects in the laboratory. This also means the amount of isolation is likely to change with changes in the environment as the reflected signal paths would change. However, the common observation of importance in both environments, is that the antenna array provides an improvement in isolation.

5.3.3 Summary

It can be seen that using an antenna array to produce a null in a particular direction and placing the receive antenna in the respective null produces more isolation between transmit and receive. Therefore, it is a suitable technique to be applied with the STAR demonstrator.

5.4 Analogue cancellation performance

The analogue cancellation performance is also measured using S-parameters. The S_{21} parameter is used to quantify the isolation between the transmit and receive channels. By combining the analogue cancellation layer with the passive suppression layer, the combined performance of the two cancellation layers can be obtained as well as the performance of the analogue cancellation layer itself, in the combined system.

5.4.1 Initial testing

In order to evaluate the analogue RF cancellation, an initial test was first performed using an attenuator to model the isolation between the transmit and receive antennas along with the analogue cancellation module that was designed.

The S-parameters of the system were measured on the VNA before cancellation (without the analogue cancellation hardware) and after applying analogue cancellation. The results are shown in Figure 5.5. A summary of the cancellation results are shown in Table 5.2.

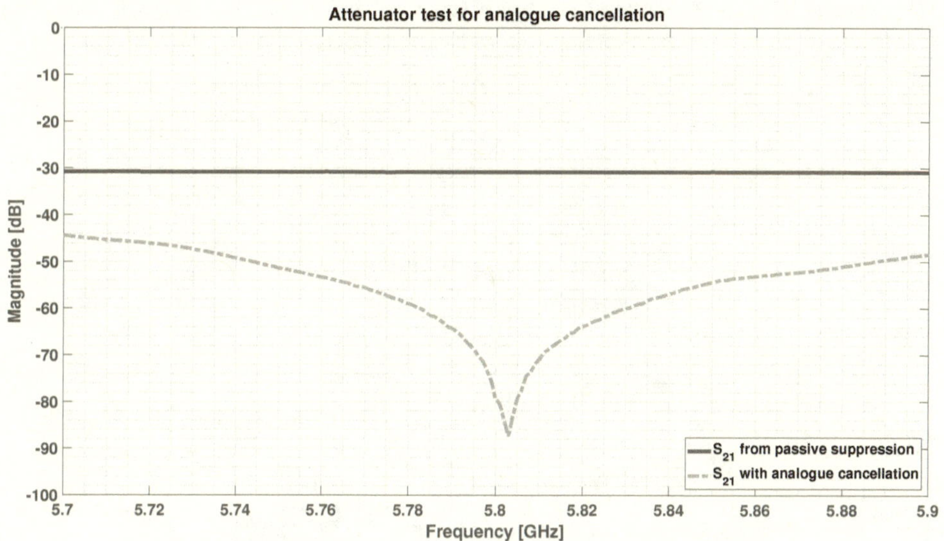

Fig. 5.5. Attenuator test for RF cancellation

Table 5.2. Summary of attenuator test for RF cancellation

Cancellation [dB]	Bandwidth of cancellation [MHz]
10	200+
20	135
30	44
40	14
50	4

From Figure 5.5 and Table 5.2 it can be seen that after inserting the analogue cancellation hardware and applying cancellation, in excess of 30 dB of cancellation is achieved over a bandwidth of 44 Mhz, 20 dB over a larger bandwidth of 135 MHz and 10 dB over 200+ MHz bandwidth, centred at 5.803 GHz. This is indicative of the canceller being able to cancel the direct signal interference very well over a very narrow bandwidth, and less effectively as the bandwidth is increased. These results are considered to be for an ideal case since passive attenuation is not fully representative of antennas which experience various EM effects.

5.4.2 Analogue RF cancellation results

The analogue RF cancellation hardware was setup as in the diagram in Figure 5.6. The VNA was then used to measure the S-parameters while applying cancellation.

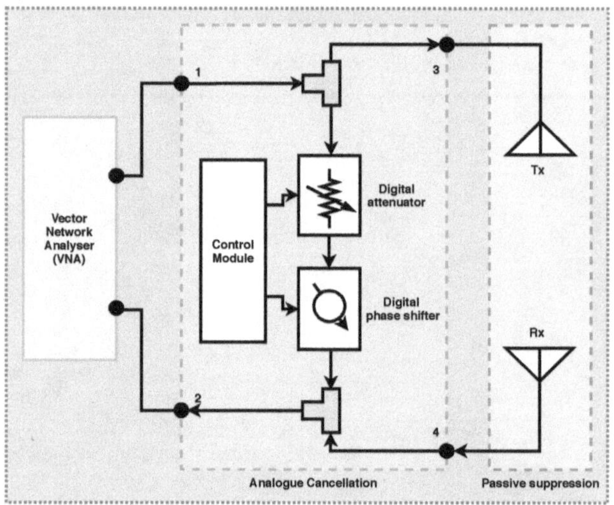

Fig. 5.6. Analogue cancellation measurement setup.

These measurements were conducted in the laboratory environment. First a single transmit antenna was tested and thereafter the linear antenna array was tested. The results presented include the measured analogue cancellation results along with the passive suppression results for comparison to see how much more cancellation is achieved.

Analogue cancellation - Single transmit antenna results

The measurement results with maximum cancellation using a single transmit antenna is presented in Figure 5.7 and summarised in Table 5.3.

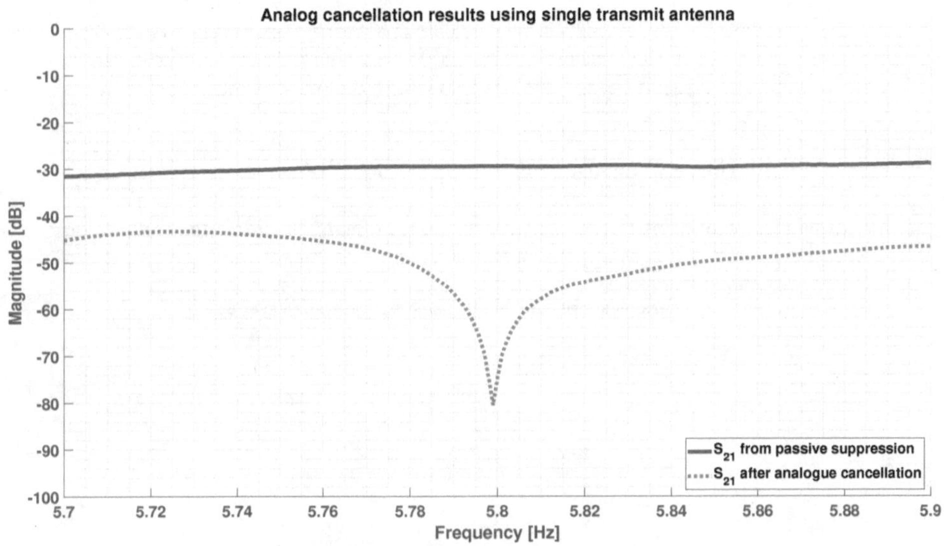

Fig. 5.7. Measured S_{21} before and after RF cancellation using single transmit antenna

Table 5.3. Summary of RF cancellation using single transmit antenna

Cancellation [dB]	**Bandwidth of cancellation [MHz] Laboratory**
10	200+
20	65
30	15

From Figure 5.7 and Table 5.3 it can be seen that significant cancellation is achieved. The notch shaped response indicates a large amount of cancellation at a single frequency, with the amount of cancellation decreasing as the bandwidth is increased. At least 10 dB of cancellation is achieved over 200+ MHz of bandwidth and at least 30 dB over 15 MHz centred at 5.8 GHz. It can also be noted that this measured result has a similar shaped S_{21} graph as that of the the attenuator test. However, the cancellation bandwidth is much lower.

The notch shaped response is due to the technique of cancellation used. The cancellation signal is generated by attenuating and phase shifting a part of the transmit signal in order to cancel out the transmit signal present in the received signal. However, the $\pi/2$ offset used to separate the transmit and receive antennas is only precise for a single frequency, in this case the centre frequency. Any bandwidth around the centre frequency will not experience the precise phase shift required to make the cancellation signal cancel perfectly with the interference signal. Hence the presence of the notch in the measured results after cancellation. This analogue cancellation can therefore be classified as a frequency dependent narrowband system.

Analogue cancellation - Transmit antenna array results

The measurement results with maximum cancellation using a transmit antenna array is presented in Figure 5.8 and summarised in Table 5.4.

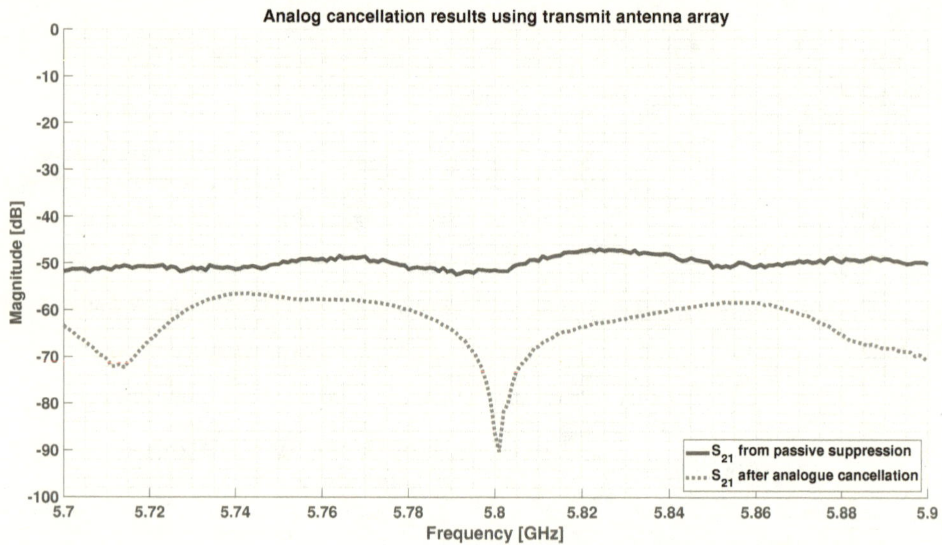

Fig. 5.8. Measured S_{21} before and after RF cancellation using transmit antenna array

Table 5.4. Summary of RF cancellation using transmit antenna array

Cancellation [dB]	**Bandwidth of cancellation [MHz]** **Laboratory**
10	45
20	10
30	3

From Figure 5.8 and Table 5.4 can be seen that there is a similar trend to the results using a single antenna. The major difference being that strong cancellation is achieved over a much narrower bandwidth. At least 10 dB of cancellation is achieved over 45 MHz of bandwidth and at least 30 dB over 3 MHz centred at 5.8 GHz.

5.4.3 Summary

The results show that the analogue RF cancellation hardware is operational and performs effective cancellation centred at 5.8 GHz. It is observed that the amount of cancellation decreases with increasing bandwidth. In terms of performance, using the antenna array results in narrower bandwidth analogue cancellation than using a single transmit antenna. However, since the focus of this project is on narrowband self-interference cancellation, these results demonstrate that the analogue cancellation is effective for narrowband self-interference cancellation.

5.5 Digital cancellation performance

5.5.1 Digital cancellation processing

Matlab implementations of the CGLS and ECA algorithms were developed for passive radar projects within the RRSG in [38]. These algorithms have been adapted to work with the STAR demonstrator setup. The details of the integration of these algorithms into the STAR demonstrator are discussed below.

5.5.2 Recording signal data

The adaptive filtering algorithms requires both a reference signal and a surveillance signal that have been coherently recorded. The surveillance signal is the signal received from the receive antenna. The reference signal is the signal that is to be removed from the surveillance signal, in this case that is the transmit signal.

The Ettus Universal Software Radio Peripheral (USRP) N210 SDR with the SBX USRP daughterboard is used to record the signal data. The SBX daughterboard is a wide bandwidth transceiver with an operational band of 400 MHz to 4400 MHz [51]. Since the hardware is not operational at 5.8 GHz, a centre frequency of 1.8 GHz is used for compatibility with the SDR. Additional hardware is required to up-convert the signal in the transmit chain and down-convert the signal in the receive chain. The frequency conversion and its analysis is discussed further in Section 5.5.3.

The Laboratory Virtual Instrument Engineering Workbench (LabVIEW) software platform was used to configure the hardware to record two signals. Coherency of the two recorded signals was achieved by using two SDRs in MIMO configuration. The recorded data was saved in a binary file format which can easily be read by the Matlab scripts for processing .

Fig. 5.9. The Ettus USRP N210 SDR.

5.5.3 Frequency up-conversion and down-conversion

The Ettus N210 SDR SBX daughter boards available operate with signals in the range of 0.4 GHz to 4.4 GHz. Since the carrier frequency being used is 5.8 Ghz and the SDRs will be operating at 1.8 GHz, frequency up-conversion and down-conversion is required. A Mini-Circuits ZX05-U742MH+ passive mixer (0.1 MHz - 7400 MHz) [52] is used for up-conversion and a Mini-Circuits ZX05-C60MH+ passive mixer (1600 MHz - 6000 MHz) [53] is used for down-conversion.

The RF frequency used on the Ettus boards is 1.8 GHz. Thus, to get an intermediate (IF) frequency of 5.8 GHz, a local oscillator (LO) frequency of 4 GHz must be used. Since a mixer is a non-linear device, the output of the mixers contain not only the desired sum and difference frequency components, but also intermodulation distortion components. The hardware configuration is shown in Figure 5.10 and a frequency chart identifying all these different frequency components is shown in Figure 5.11.

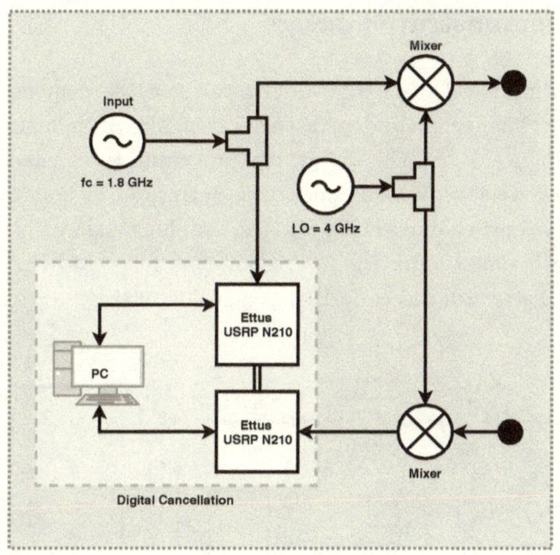

Fig. 5.10. Illustration of the setup including mixers and signals generators.

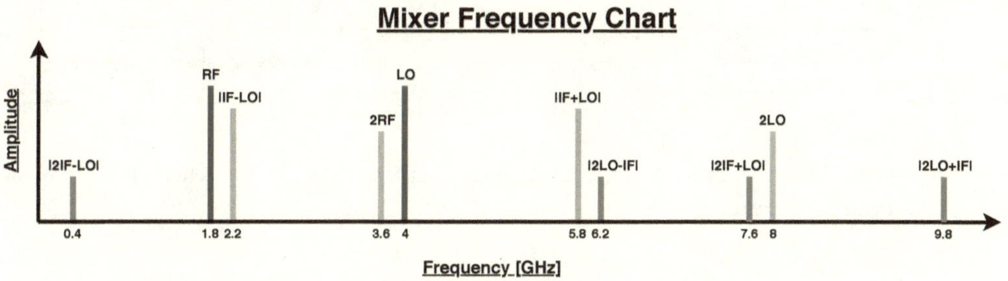

Fig. 5.11. Frequency chart identifying all the frequency components present in the system when mixers are used.

Good practice for both the transmit and receive channels is to use filters to remove all the unwanted components. The transmit channel must be filtered to avoid transmitting at frequencies that fall outside the unlicensed ISM band. The receive channel must be filtered to reduce noise and interference in the receive channel before amplification as well as to filter out the image frequency so that it does not mask the desired received signal after frequency down-conversion.

5.5.4 Complete demonstrator design

In order to perform digital cancellation, the all the parts of the demonstrator needs to be integrated and functional. The complete demonstrator design with its multi-layer cancellation scheme is presented in Figure 5.12. The design comprises the three cancellation techniques previously discussed along with additional supporting hardware. A band-pass filter is added after the mixer in the transmit channel along with an amplifier before the transmit antenna. On receive, another filter is placed after the receive antenna followed by a low noise amplifier (LNA) to boost the signal strength before further processing in the receive channel.

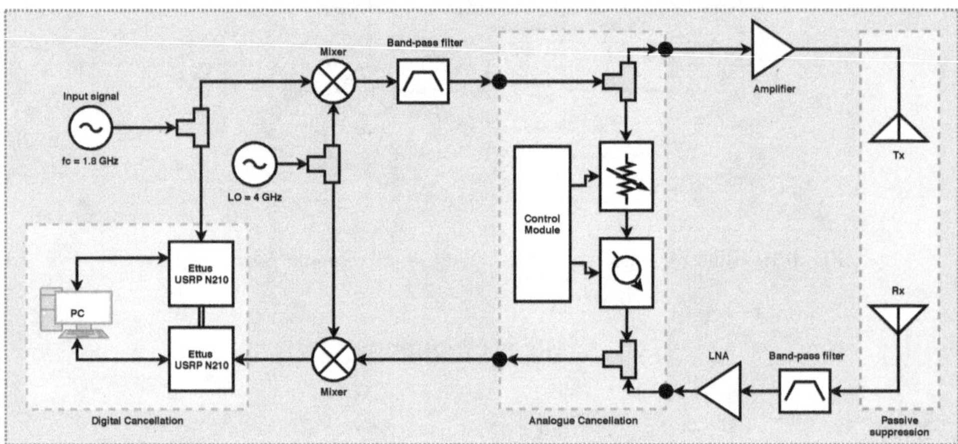

Fig. 5.12. The complete demonstrator design with multi-layer cancellation scheme and supporting hardware.

5.5.5 Supporting hardware

Due to financial constraints it was not possible to acquire all the supporting hardware. The LNA and a second filter are therefore absent and the complete demonstrator had to therefore be constructed without these. The Mini-Circuits ZX60-83LN-S+ LNA [54] with a gain of approximately 20 dB would have been used and a suitable microstrip band-pass filter would have been designed and manufactured if funds were available. Therefore, the constructed demonstrator is illustrated in Figure 5.13.

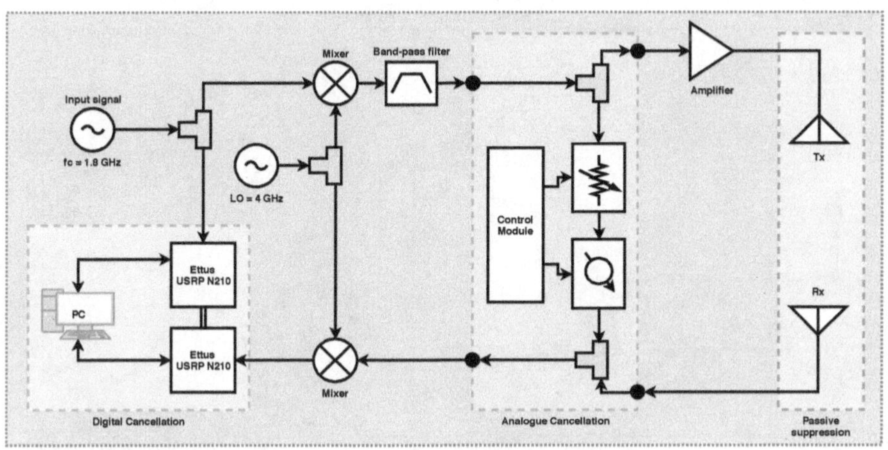

Fig. 5.13. The constructed demonstrator design with multi-layer cancellation scheme and supporting hardware, without filter and LNA on receive.

The amplifier which is used is a Mini-Circuits ZX60-V63+ wideband amplifier [55] which operates from 50 MHz to 6 GHz with a gain of approximately 15 dB at 5.8 GHz. The band-pass filter which is used is a microstrip hairpin filter that was available in the lab. It was measured using the VNA to obtain the measured filter response. An image of the filter is shown in Figure 5.14 and the measured S-parameters of the filter are presented in Figure 5.15. The filter response indicates an insertion loss of approximately 1.7 dB with a pass-band from 4.88 GHz to 6.48 GHz (1.6 GHz bandwidth).

Fig. 5.14. Image of the microstrip hairpin filter.

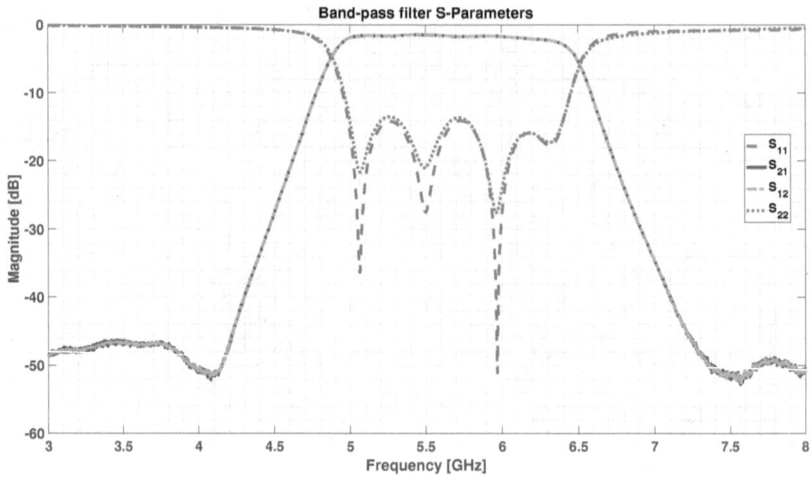

Fig. 5.15. Measured S-parameters of band-pass filter.

5.5.6 Transmit signal

A 4 MHz chirp modulated signal is used to assess the performance of the system. The bandwidth is the parameter of interest. The period and pulse width parameters were selected arbitrarily, the aim being to have a signal with some bandwidth with which to test the system performance. The signal parameters used are detailed in Table 5.5. A normalised spectrum of the chirp signal can be found in Figure 5.16.

Table 5.5. Chirp modulation signal parameters

Parameter	**Value**
Centre frequency	1.8 GHz
Bandwidth	4 MHz
Pulse period	100 μs
Pulse width	50 μs
Total signal power	20 dBm

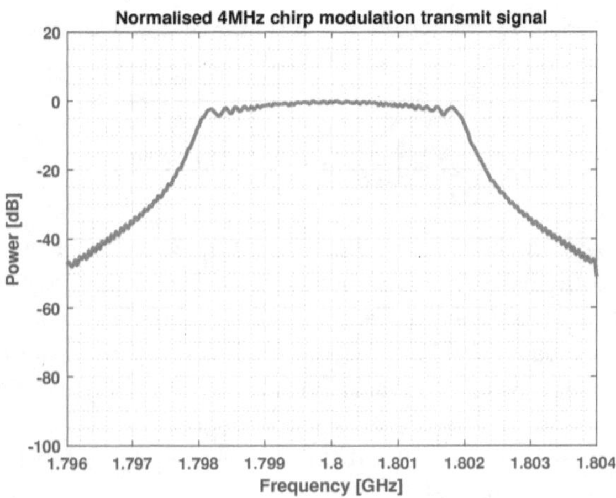

Fig. 5.16. Frequency spectrum of the 4MHz chirp modulated signal centred on 1.8 GHz.

5.5.7 Setup and measurement procedure

The digital cancellation is done by recording data using the SDRs and then processing the data using the Matlab implementations of the CGLS and ECA algorithms. As discussed in Section 5.5.3, frequency heterodyning is needed to be able to operate within the working frequency range of the SDRs. These experiments were performed in the multipath laboratory environment.

The input signals are generated using the signal generator (4 MHz chirp centred at 1.8 GHz) and sweep oscillator (4 GHz). The analogue cancellation may require additional attenuators to generate the cancellation signal if the amount of passive suppression from the antennas is high. If there is insufficient attenuation, it will not be possible to match the amplitude of the cancellation signal with that of the self-interference.

The demonstrator was setup as shown in Figure 5.17 using a single transmit antenna. Before recording the signals using the SDRs, the signal power was measured at various points in the system using a spectrum analyser to confirm that the system functions as expected and determine if there are any problems within the system. For these measurements the analogue cancellation hardware was removed from the system and the connection points thereof terminated in 50 ohm loads.

The spectrum analyser is configured to have a resolution bandwidth (RBW) of 1 kHz over a 20 MHz span with 400 data points. It should be noted that the spectrum analyser does not measure the total power of the signal, but rather the amount of power contained within a specific frequency band set by the RBW. The total signal power is computed by the following formula [56]:

$$P_{ch} = \left(\frac{B_s}{B_n}\right)\left(\frac{1}{N}\right)\sum_{i=n1}^{n2} 10^{(p_i/10)} \tag{5.1}$$

where P_{ch} is the power in the channel (in milliwatts), B_s is the channel bandwidth, B_n is the RBW used, N is the number of data points in the summation, p_i is the sample of power in measurement cell i in dBm, and $n1$ and $n2$ are the end-points for the index i, i.e. $N = (n2 - n1) + 1$. An example of calculating the total signal power from a spectrum analyser measurement is given in Appendix A. From Appendix A, the total signal power is approximately 30 dB more than the measured power in a single frequency bin (at the centre of the chirp spectrum) on the spectrum analyser. Therefore, all the measured power levels are increased by 30 dB to get the total signal power.

The test points are indicated by the letters a to k in Figure 5.17. After all the test points were measured, this was repeated for the system using the transmit antenna array. The total signal power has been computed for each test point using Equation 5.1 and are given in Table 5.6.

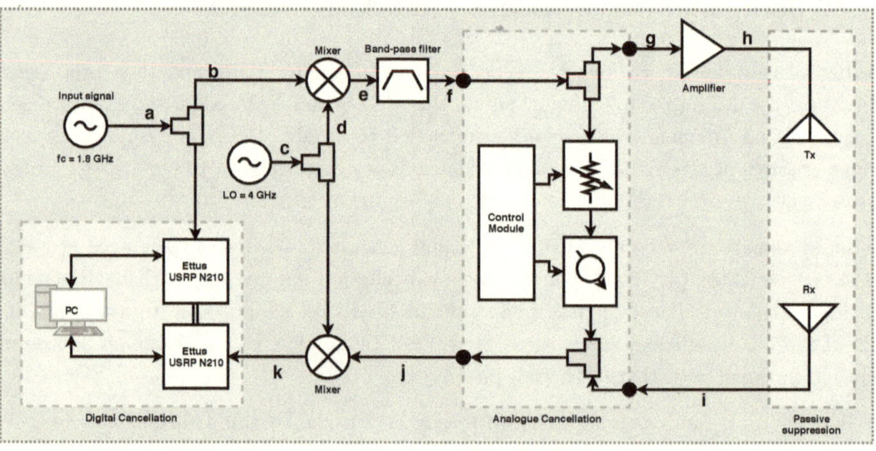

Fig. 5.17. The constructed demonstrator design with multi-layer cancellation scheme and supporting hardware, without filter and LNA on receive. Spectrum analyser test points are indicated by the letter a to k.

Table 5.6. Total signal power levels at various test points in the system.

Test point	Total signal power [dBm]	
	Single Tx antenna	Tx antenna array
a	20	20
b	17	17
c	17	17
d	13	13
e	5	5
f	2	2
g	-1	-1
h	12	12
i	-22	-46
j	-25	-49
k	-27	-27

From the measured power levels it can be seen that in the demonstrator, the up-conversion mixer has a conversion loss of 12 dB, the filter has an insertion loss of 3 dB and the amplifier provides a gain of 13 dB. However, the down-conversion mixer appears to provide a conversion gain instead of a conversion loss. Also, the power level after down-conversion is the same for both the single transmit antenna as well as the transmit array. This indicates that there is an issue within the system.

Further measurements were taken to find the problem. This included measuring the leakage signals through the mixing stages. It was discovered that the leakage signal from the transmitter through the mixing stages was stronger than the received signal which is being down-converted. Hence the received signal was being masked by the leakage signal, resulting in a higher signal power than expected.

Increasing the received signal power would require the addition one or more amplification stages in the system. But as previously mentioned, additional hardware could not be acquired due to financial constraints. Therefore, the digital cancellation could not be properly tested while integrated with the passive suppression and analogue cancellation layers. Instead, the digital cancellation was tested by simulating 20 dB of amplification gain on receive.

Passive suppression using a single transmit antenna provides approximately 30 dB of self-interference cancellation. 20 dB of gain is simulated by using a 10 dB passive attenuator in place of the transmit and receive antennas. This ensures that the received signal is stronger than the leakage signal and will provide better signal integrity for the digital cancellation. An experiment is then performed by adding analogue cancellation as well. However, when using analogue cancellation, the received signal will likely be weaker than the leakage signal and affect the integrity of the results. With this in mind, the experiment is still performed to analyse the effect of digital cancellation.

The two Ettus SDR boards are setup in MIMO configuration with two receive channels, one for the reference (transmit) signal and one for the surveillance (received) signal. Each of the channels are configured to have a gain of 10 dB. The centre frequency is set to 1.8 GHz and the in phase and quadrature (IQ) sampling frequency of the signals is 8 MHz. A total of ten seconds of data is recorded.

5.5.8 Digital cancellation results

First the received signal without analogue cancellation was recorded and processed. The digital cancellation results are shown in Figures 5.18 (using CGLS method) and 5.19 (using ECA method). Only the envelopes of the frequency spectrum of the signals are plotted for better visualisation. The plots are normalised to the first measured received signal when no analogue or digital cancellation was applied.

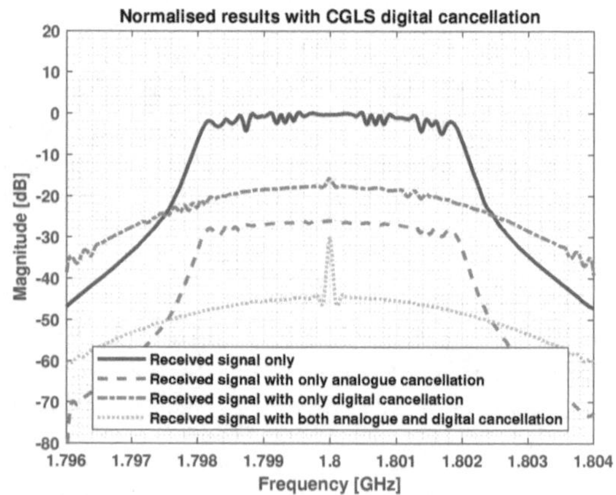

Fig. 5.18. Digital cancellation results using CGLS algorithm.

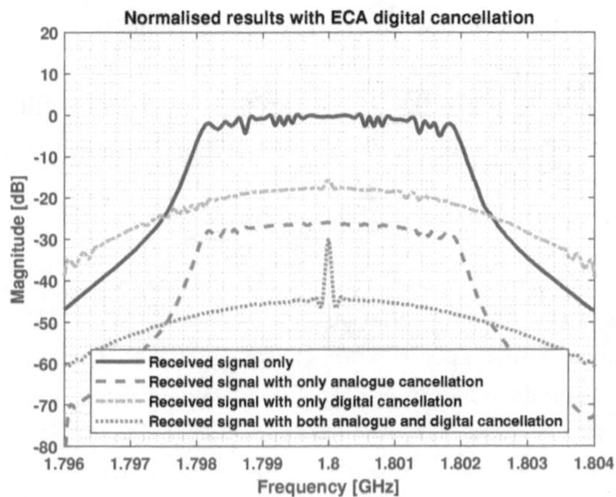

Fig. 5.19. Digital cancellation results using ECA algorithm.

From Figure 5.18 it can be seen that the experiment which applies only digital cancellation achieves 16 dB of cancellation over the bandwidth of the signal.

The experiment with analogue cancellation records 26 dB of analogue cancellation. However, this is much less than what was expected. According to the initial analogue cancellation VNA test using an attenuator for passive suppression, there should be in excess of 30 dB, up to 50 dB of analogue cancellation over a 4 MHz bandwidth. The measured signal power (using a spectrum analyser) showed -60 dBm before down-conversion (test point k), which means that it was weaker than the leakage signal. This means that the leakage signal is dominant and the signal that is recorded is not of high fidelity and will not give a proper representation of the results of cancellation.

After applying digital cancellation to the signal which underwent analogue cancellation, an additional 20 dB of cancellation is achieved, with the exception of a spike at 1.8 GHz. This spike is local oscillator leakage from the transmit channel. The digital cancellation uses a chirp signal as the reference and thus is able to suppress a chirp signal. However, it cannot suppress any leakage signals because these are not captured in the reference channel.

The ECA digital cancellation results in Figure 5.19 look almost identical to the CGLS results. This can be attributed to the fact that although the two methods use a different approach (CGLS is an iterative method, ECA is not iterative), they are both least squares algorithms, hence they produce similar results. This is a useful observation because, depending on the application timing requirements and computing resources available, one could choose between the two algorithms with minimal difference in the result.

These results show that digital cancellation does work and provides a significant amount of additional cancellation. However, due to the low power levels in the system, the integrity of the received signal which undergoes cancellation is affected by the unwanted leakage signal in the system. The solution is to make the robust through the use of additional hardware such as filters and amplifiers in various parts of the system to help maintain signal integrity.

5.5.9 Summary

The digital cancellation results show that an additional 16 to 20 dB of cancellation can be achieved from applying digital cancellation. It was identified that the missing amplifier on receive is important to maintain the integrity of the received signal and avoid it being masked by leakage signals in the system. Greater care can also be taken to prevent the leakage by adding more filters in the system. It was also observed that the ECA and CGLS algorithms both produced similar results.

5.6 Further analysis of results

5.6.1 Frequency selectivity, bandwidth limitations

The system performance points towards a frequency selective system. The amount of cancellation achieved is dependent on frequency. There is also a bandwidth limitation which stems from this. High amounts of cancellation can be achieved, but over very narrow bandwidths. However, this is limited by hardware precision and accuracy as well as uncontrollable effects such as multipath.

5.6.2 Hardware precision

Hardware precision encompasses hardware limitations, manufacturing tolerances, operating error as well as geometric placement errors. These errors make it difficult to achieve perfect self-interference cancellation. Dealing with these issues is not critical to the goal of the project. However, mitigating these factors and reducing their effects are important if the system were to be optimised.

5.6.3 Multipath effects

Multipath plays an important role in any system that uses wireless RF signals, the STAR demonstrator included. The operational environment typically contains both static (from stationary objects) and dynamic (from moving objects) multipath. Simply changing the position of an object within an environment can change the multipath reflections within that environment. A moving object within an environment will cause a dynamic change in the multipath reflections. In order to compensate for this in the analogue domain, an adaptive system is required.

The STAR demonstrator with statically controlled analogue cancellation cannot adapt itself to compensate for any changes in multipath. Thus it can essentially cancel a particular instance of interference very well, but any change would affect performance of the system and require reconfiguration of the cancellation signal. This is effective mainly for suppressing self-interference, but not multipath. Multipath can however be dealt with to some extent in the digital domain. Algorithms such as CGLS and ECA can be configured to support changes in position and Doppler. Anything outside the configured parameters will not be suppressed by digital cancellation.

5.7 Overall system performance summary

An overall summary of the complete STAR demonstrator with self-interference cancellation is presented in Table 5.7. The performance of each technique is quantified as part of the fully integrated demonstrator. Whilst analogue cancellation could not be properly quantified after digitisation, it is expected that the results from the analogue cancellation testing will still apply as digitisation should not have any major effects on it.

Table 5.7. Performance summary

Technique	**Performance [dB]**
Passive suppression	29.4 to 51.5
Analogue cancellation	20 to 30
Digital cancellation	16 to 20
Total	**65.4 to 101.5**

The first technique, passive suppression, using a dual antenna setup provides 29.4 dB of isolation. When using a transmit array the amount of isolation increases by 22.1 dB to 51.5 dB. This shows that using an antenna array is an effective method of improving transmit-receive isolation. The second technique, analogue cancellation, is capable of of producing 20 to 30 dB of cancellation when used with the antenna array. This plays a significant role in the system to help prevent saturating the receiver. And the third technique, digital cancellation, provides 20 dB of additional cancellation. Collectively, 65.4 to 101.5 dB of cancellation is obtainable, thereby indicating that the techniques which were used are effective on a narrowband signal.

Chapter 6

Conclusion and Recommendations for Future Work

6.1 Conclusions

As frequency spectrum is becoming increasingly limited and overcrowded, it is driving research in STAR systems, which are able to transmit and receive on the same frequency at the same time. In other words, there is a growing need for systems with higher spectral efficiency. The challenge with STAR systems is that they require high isolation between transmit and receive channels to overcome the obstacle of self-interference. A lack of isolation will result in a significant reduction in the receiver sensitivity and dynamic range, reducing its ability to adequately detect incoming signals. Therefore, the suppression of this self-interference is vital to enabling STAR capability.

The focus of this study was to evaluate the theoretical and practical viability of a STAR system with co-located transmit and receive channels. This was done through an investigation into self-interference suppression techniques followed by the development of a narrowband self-interference cancellation demonstrator. The approach to the demonstrator entails using a multi-layer cancellation scheme comprising of passive suppression, RF cancellation and digital cancellation techniques. It is built around WiFi antennas which operate at the 5.8 GHz ISM frequency band.

The passive suppression layer employs separate transmit and receive antennas which are co-located. In order to improve the isolation between transmit and receive, a four element linear antenna array with half a wavelength spacing between elements is utilised to create a directional transmitter, where the receiver is co-located in a null of the transmitter, resulting in increased isolation.

The analogue cancellation requires splitting a signal from the transmit chain and applying an attenuation and phase shift to this signal, then adding it to the received signal. The HMC1122LP4ME 6-bit digital step attenuator and HMC649ALP6E 6-bit digital phase shifter from Analog Devices was used to construct the analogue cancellation circuitry. PCBs were designed for these devices, along with a mechanical switch based control module to interface with the devices. Since the main objective is not to develop an optimal system, the analogue cancellation is not made to be adaptive.

For the digital cancellation implementation, existing CGLS and ECA algorithms have been adapted to form the third layer of cancellation in the system. The cancellation algorithms were executed using Matlab on a desktop computer after digitising and recording the signal data.

The transmit-receive isolation between the antennas is obtained from the transmission coefficient S_{21}. The transmit-receive isolation was measured both in an anechoic chamber as well as in a multipath laboratory environment, and testing was done using both a single transmit antenna and the transmit antenna array. In an anechoic chamber, the single transmit antenna produced 44 dB of isolation and the antenna array produced 58 dB of isolation. This indicates an improvement of 14 dB when using the antenna array. When tested in a more realistic multipath environment, the array showed an improvement in isolation of 22 dB, from 29 dB isolation (single transmit antenna) to 51 dB isolation (transmit array). This demonstrates that using an antenna array is an effective passive suppression technique. The main caveat with this technique is that the multipath environment may change and affect the isolation.

The analogue RF cancellation hardware was set up for experimentation with the transmit antenna array and receive antenna to measure the standalone performance of the analogue cancellation hardware. A vector network analyser (VNA) was used to measure the S-parameters before and after applying analogue cancellation in a multipath environment. The S_{21} after analogue cancellation demonstrates 10 dB cancellation over a 40 MHz bandwidth and in excess of 40 dB of cancellation as the bandwidth is decreased around the centre frequency of 5.8 GHz. However, this is a standalone best case scenario and does not necessarily represent performance when integrated with other techniques and hardware.

Digital cancellation requires the digitisation of signals before it can be applied. For the CGLS and ECA methods, both the transmit and received signals need to be coherently sampled. The Ettus USRP N210 SDR with the SBX transceiver daughterboard is the available hardware which was used to digitise and record the signals. However, the SBX transceiver only operates from 400 Mhz to 4.4 GHz, thus frequency up-conversion and down-conversion is needed on the transmit and receive channels respectively, for compatibility with the antennas and analogue cancellation hardware which operate at the 5.8 GHz frequency band. Two SDRs set up in multiple-input multiple-output (MIMO) configuration are used to obtain signal coherency. Experiments were then carried out using a 4 MHz chirp transmit signal in order to quantify the performance of analogue cancellation (as part of the larger system) and digital cancellation.

Digital cancellation was performed using both the CGLS and ECA methods for comparison. The normalised received signal after passive suppression is used as the benchmark to be able to quantify the analogue and digital cancellation performance. A total of 26 dB of analogue cancellation was achieved. Digital cancellation produced 16 to 20 dB of cancellation. Overall, more than 45 dB of cancellation was achieved when combining analogue and digital cancellation in the demonstrator. When comparing the two digital cancellation methods, they produce similar results. The reason for this being that though they are different methods, both the CGLS and ECA methods are fundamentally least squares based algorithms.

From the results achieved, it is clear that passive suppression provides the largest amount of isolation. Commercial WiFi antennas were used because of their cost and availability, and were not expected to produce the best possible results. However, they were still able to produce significant isolation of 29.4 dB between transmit and receive and were thus suitable for such an experimental system. An equally phased linear antenna array was chosen as a means to improve the isolation, because it is not complex to construct and requires minimal extra hardware. This ended up as a good design decision and provided a 22 dB improvement in isolation.

Analogue cancellation has proved useful in supplementing passive suppression to significantly reduce self-interference. Even though the system was not adaptive, it showed significant narrowband cancellation capability. It should be noted though, that the system was not tested in a dynamic multipath system where adaptive capability would be important. The hardware used for the analogue cancellation is not suited for wideband cancellation. But alternative hardware could enable wideband cancellation.

Digital cancellation, when used in conjunction with the aforementioned techniques, is effective in suppressing any residual self-interference. It even manages to suppress the leakage signal. The presence of the leakage signal signifies the need for refinement and more careful design to ensure it does not affect the integrity of signals in the system. One method of handling this is to include more filters in the signal paths around the mixers.

A summary of the overall performance of the cancellation techniques used in the demonstrator is presented in Table 6.1. A total of 65.4 to 101.5 dB of cancellation is achievable from passive suppression, analogue cancellation, and digital cancellation together in the demonstrator. This shows that each layer of cancellation in the demonstrator is effective. Typically, one would require in excess of 90 dB of self-interference cancellation in a STAR system. The results of this demonstrator indicate that it is capable of achieving in excess of 90 dB of self-interference cancellation. For an experimental system, the results look promising, but the demonstrator has much room for improvement and optimisation, which should improve performance even more. Overall, it can be concluded that STAR systems are practically viable as achieving sufficient self-interference cancellation is possible.

Table 6.1. Performance summary

Technique	**Performance [dB]**
Passive suppression	29.4 to 51.5
Analogue cancellation	20 to 30
Digital cancellation	16 to 20
Total	**65.4 to 101.5**

6.2 Recommendations for future work

Considering that the system is experimental and not optimised, it should be possible to attain more isolation with various improvements and optimisation. Below are some recommendations for future work to improve and optimise the demonstrator.

Adaptive analogue cancellation

The demonstrator which was developed does not use adaptive analogue cancellation and can therefore not respond to the multipath effects of a changing environment. By making the system adaptive, it will be able to dynamically adjust the attenuation and phase shift to achieve the best possible cancellation. This would require a feedback loop with amplitude and phase detection along with an embedded system to compute and provide control inputs to the analogue cancellation system.

Design high isolation transmit-receive antenna pair

Ideally, a high isolation transmit-receive antenna pair should be designed in order to get the maximum amount of passive suppression. The exact specifications of such a design would depend on the application requirements. For example, in a high power CW radar system, a specialised horn antenna could be designed, for automotive radar, a microstrip antenna and for a line of sight communications link, a reflector antenna. Ultimately, the specifications such as size, bandwidth, polarisation and form factor all depend on the application for which it is being used.

Improve analogue cancellation hardware

The performance of a system generally is limited by the hardware in the system. The capability of the hardware, operating ranges, tolerances and losses all play a role in the overall system performance. The prototype digital attenuator and phase shifter modules which were designed and manufactured have quite high insertion losses due to impedance mismatches. This could be improved by using stepped impedance transformers for proper impedance matching or purchasing an evaluation board. One could also use components with higher resolution and lower tolerances to improve the accuracy of the system.

Only a single cancellation tap was used to target cancellation of the self-interference signal. This can be improved upon by using multiple tapped lines with various delays. This can help cancel out some of the multipath interference signals.

An alternative hardware configuration can also be used such as using a balun in place of the phase shifter for generating the cancellation signal. This should have the added benefit of enabling cancellation over a wider bandwidth. Another suitable alternative would be to use a vector modulator which can do both the amplitude and phase adjustment.

More comprehensive front end design

The demonstrator designed is an experimental system and the RF front end has not been comprehensively designed. Apart from the components missing due to budget limitations, further amplification stages and filtering can be added into the system. Also, alternative methods can be explored for generating the transmit signal.

Noise analysis

This study does not analyse or quantify the noise in the system. The analogue cancellation hardware introduces noise into the system which can affect performance. For future work on this project it is recommended that an in depth analysis be done on noise in the system.

Appendix A

Total signal power from spectrum analyser measurements

Presented below is an example of calculating the total signal power from the measured signal power on a spectrum analyser. The spectrum analyser is configured to have a resolution bandwidth (RBW) of 1 kHz over a 20 MHz span with 400 data points.

A 4 MHz chirp signal centred at 1.8 GHz was generated using the signal generator. The total power output from the signal generator was set to 0 dBm (1 mW). The output signal from the signal generator was then connected with a cable to the spectrum analyser and measured. The following spectrum in Figure A.1 was obtained.

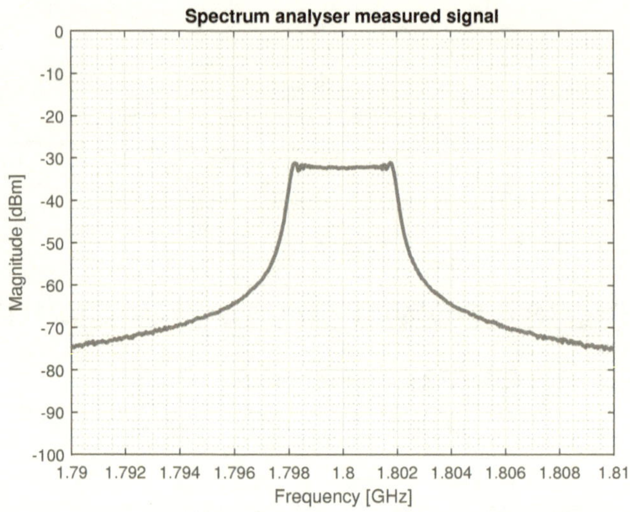

Fig. A.1. Measured signal from spectrum analyser.

The measured chirp signal power is approximately -32 dBm at 1.8 GHz. In order to compute the total signal power, the spectrum analyser signal was saved in CSV format and imported into Matlab.

According to Equation 5.1, the total power in the signal is the sum of the power in each of the frequency bins over the bandwidth, multiplied by the scale factor $\frac{B_s}{B_n N}$. For this calculation, $B_s = 4\ MHz$, $B_n = RBW = 1\ kHz$ and $N = 401$. If the signal begins with noise like characteristics, it will incur a 2.5 dB under-response (refer to [56] for more details). Thus 2.5 dB needs to be added to the result to get an accurate measurement.

The result of applying Equation 5.1 is -3.2 dBm. Since the signal starts in the noise floor, it is deemed to have noise like characteristics and requires and additional 2.5 dB to be added. Therefore the final result for the total signal power is -0.7 dBm, which is fairly close to the 0 dBm that was output from the signal generator. The remaining 0.7 dB difference is due to cable loss.

References

[1] K. Kolodziej, P. Hurst, A. Fenn, and L. Parad, "Ring array antenna with optimized beamformer for Simultaneous Transmit And Receive," in *Proceedings of the 2012 IEEE International Symposium on Antennas and Propagation*, pp. 1–2, IEEE, 7 2012.

[2] K. L. Scherer, S. J. Watt, E. A. Alwan, A. A. Akhiyat, B. Dupaix, W. Khalil, and J. L. Volakis, "Simultaneous transmit and receive system architecture with four stages of cancellation," in *2015 IEEE International Symposium on Antennas and Propagation & USNC/URSI National Radio Science Meeting*, pp. 520–521, IEEE, 7 2015.

[3] A. Sabharwal, P. Schniter, D. Guo, D. W. Bliss, S. Rangarajan, and R. Wichman, "In-Band Full-Duplex Wireless: Challenges and Opportunities," *IEEE Journal on Selected Areas in Communications*, vol. 32, pp. 1637–1652, 9 2014.

[4] K. E. Kolodziej, B. T. Perry, and J. S. Herd, "Simultaneous Transmit and Receive (STAR) system architecture using multiple analog cancellation layers," in *2015 IEEE MTT-S International Microwave Symposium*, pp. 1–4, IEEE, 5 2015.

[5] G. Liu, F. Yu, H. Ji, V. Leung, X. L. Resource, and u. 2015, "In-band full-duplex relaying: A survey, research issues and challenges," *ieeexplore.ieee.org*.

[6] Z. Zhang, K. Long, A. V. Vasilakos, and L. Hanzo, "Full-Duplex Wireless Communications: Challenges, Solutions, and Future Research Directions," *Proceedings of the IEEE*, vol. 104, pp. 1369–1409, 7 2016.

[7] A. Goldsmith, *Wireless communications*. Cambridge University Press, 2005.

[8] S. Hong, J. Brand, J. Choi, M. Jain, J. Mehlman, S. Katti, and P. Levis, "Applications of self-interference cancellation in 5G and beyond," *IEEE Communications Magazine*, vol. 52, pp. 114–121, 2 2014.

[9] M. Duarte and A. Sabharwal, "Full-duplex wireless communications using off-the-shelf radios: Feasibility and first results," in *2010 Conference Record of the Forty Fourth Asilomar Conference on Signals, Systems and Computers*, pp. 1558–1562, IEEE, 11 2010.

[10] J. I. Choi, M. Jain, K. Srinivasan, P. Levis, and S. Katti, "Achieving single channel, full duplex wireless communication," in *Proceedings of the sixteenth annual international*

conference on Mobile computing and networking - MobiCom '10, (New York, New York, USA), p. 1, ACM Press, 2010.

[11] M. Jain, J. I. Choi, M. Kim, D. Bharadia, S. Seth, K. Srinivasan, P. Levis, S. Katti, and P. Sinha, *Practical, Real-time, Full Duplex Wireless.* 2011.

[12] D. Bharadia, E. McMilin, S. Katti, D. Bharadia, E. McMilin, and S. Katti, "Full duplex radios," in *Proceedings of the ACM SIGCOMM 2013 conference on SIGCOMM - SIGCOMM '13*, vol. 43, (New York, New York, USA), p. 375, ACM Press, 2013.

[13] K. Kolodziej, J. McMichael, and B. Perry, "Adaptive RF canceller for transmit-receive isolation improvement," in *2014 IEEE Radio and Wireless Symposium (RWS)*, pp. 172–174, IEEE, 1 2014.

[14] M. Heino, D. Korpi, T. Huusari, E. Antonio-Rodriguez, S. Venkatasubramanian, T. Riihonen, L. Anttila, C. Icheln, K. Haneda, R. Wichman, and M. Valkama, "Recent advances in antenna design and interference cancellation algorithms for in-band full duplex relays," *IEEE Communications Magazine*, vol. 53, pp. 91–101, 5 2015.

[15] Y.-S. Choi and H. Shirani-Mehr, "Simultaneous Transmission and Reception: Algorithm, Design and System Level Performance," *IEEE Transactions on Wireless Communications*, vol. 12, pp. 5992–6010, 12 2013.

[16] E. Everett, A. Sahai, and A. Sabharwal, "Passive Self-Interference Suppression for Full-Duplex Infrastructure Nodes," *IEEE Transactions on Wireless Communications*, vol. 13, pp. 680–694, 2 2014.

[17] M. A. Khojastepour, K. Sundaresan, S. Rangarajan, X. Zhang, and S. Barghi, "The case for antenna cancellation for scalable full-duplex wireless communications," in *Proceedings of the 10th ACM Workshop on Hot Topics in Networks - HotNets '11*, (New York, New York, USA), pp. 1–6, ACM Press, 2011.

[18] W. F. Moulder, B. T. Perry, and J. S. Herd, "Wideband antenna array for Simultaneous Transmit and Receive (STAR) applications," in *2014 IEEE Antennas and Propagation Society International Symposium (APSURSI)*, pp. 243–244, IEEE, 7 2014.

[19] M. A. Elmansouri, A. J. Kee, and D. S. Filipovic, "Wideband Antenna Array for Simultaneous Transmit and Receive (STAR) Applications," *IEEE Antennas and Wireless Propagation Letters*, vol. 16, pp. 1277–1280, 2017.

[20] M. Elmansouri, P. Valaleprasannakumar, E. Tianang, E. Etelisi, and D. Filipovic, "9-1: 0.5-45GHz Simultaneous Transmit and Receive (STAR) Antenna System for Electronic Attack," tech. rep., 2016.

[21] M. A. Elmansouri, J. Ha, and D. Filipovic, "Multioctave antenna array for simultaneous transmit and receive applications," in *2018 International Applied Computational Electromagnetics Society Symposium (ACES)*, pp. 1–2, IEEE, 3 2018.

[22] M. Elmansouri, P. Valaleprasannakumar, E. Tianang, E. Etellisi, and D. Filipovic, "Single and dual-polarized wideband simultaneous transmit and receive antenna system," in *2017 IEEE International Symposium on Antennas and Propagation & USNC/URSI National Radio Science Meeting*, pp. 1105–1106, IEEE, 7 2017.

[23] D. S. Filipovic, M. Elmansouri, and E. A. Etellisi, "On wideband simultaneous transmit and receive (STAR) with a single aperture," in *2016 IEEE International Symposium on Antennas and Propagation (APSURSI)*, pp. 1075–1076, IEEE, 6 2016.

[24] E. A. Etellisi, M. A. Elmansouri, and D. S. Filipovic, "Wideband Monostatic Simultaneous Transmit and Receive (STAR) Antenna," *IEEE Transactions on Antennas and Propagation*, vol. 64, pp. 6–15, 1 2016.

[25] E. A. Etellisi, M. A. Elmansouri, and D. S. Filipovic, "Wideband dual-mode monostatic simultaneous transmit and receive antenna system," in *2016 IEEE International Symposium on Antennas and Propagation (APSURSI)*, pp. 1821–1822, IEEE, 6 2016.

[26] J. Ha, M. A. Elmansouri, P. Valale Prasannakumar, and D. S. Filipovic, "Monostatic Co-Polarized Full-Duplex Antenna With Left- or Right-Hand Circular Polarization," *IEEE Transactions on Antennas and Propagation*, vol. 65, pp. 5103–5111, 10 2017.

[27] P. Valale Prasannakumar, M. A. Elmansouri, and D. S. Filipovic, "Broadband Reflector Antenna With High Isolation Feed for Full-Duplex Applications," *IEEE Transactions on Antennas and Propagation*, vol. 66, pp. 2281–2290, 5 2018.

[28] E. A. Etellisi, M. A. Elmansouri, and D. S. Filipovic, "Wideband monostatic spiral array for full-duplex applications," in *2017 IEEE International Symposium on Antennas and Propagation & USNC/URSI National Radio Science Meeting*, pp. 1101–1102, IEEE, 7 2017.

[29] T. Snow, C. Fulton, and W. J. Chappell, "TransmitReceive Duplexing Using Digital Beamforming System to Cancel Self-Interference," *IEEE Transactions on Microwave Theory and Techniques*, vol. 59, pp. 3494–3503, 12 2011.

[30] A. H. Abdelrahman and D. S. Filipovic, "Antenna System for Full-Duplex Operation of Handheld Radios," *IEEE Transactions on Antennas and Propagation*, pp. 1–1, 2018.

[31] S. E. Johnston and P. D. Fiore, "Full-duplex communication via adaptive nulling," in *2013 Asilomar Conference on Signals, Systems and Computers*, pp. 1628–1631, IEEE, 11 2013.

[32] M. Duarte, C. Dick, and A. Sabharwal, "Experiment-Driven Characterization of Full-Duplex Wireless Systems," *IEEE Transactions on Wireless Communications*, vol. 11, pp. 4296–4307, 12 2012.

[33] E. Everett, A. Sahai, and A. Sabharwal, "Passive Self-Interference Suppression for Full-Duplex Infrastructure Nodes," *IEEE Transactions on Wireless Communications*, vol. 13, pp. 680–694, 2 2014.

[34] M. A. Khojastepour and S. Rangarajan, "Wideband digital cancellation for full-duplex communications," in *2012 Conference Record of the Forty Sixth Asilomar Conference on Signals, Systems and Computers (ASILOMAR)*, pp. 1300–1304, IEEE, 11 2012.

[35] D. Bharadia, E. McMilin, S. Katti, D. Bharadia, E. McMilin, and S. Katti, "Full duplex radios," in *Proceedings of the ACM SIGCOMM 2013 conference on SIGCOMM - SIGCOMM '13*, vol. 43, (New York, New York, USA), p. 375, ACM Press, 2013.

[36] W. L. Stutzman and G. A. Thiele, *Antenna theory and design.* Wiley, 2013.

[37] C. A. Balanis, *Antenna theory : analysis and design.* Wiley Interscience, 2005.

[38] C. A. Tong, *A scalable real-time processing chain for radar exploiting illuminators of opportunity.* PhD book, University of Cape Town, 2014.

[39] F. Colone, D. W. O'Hagan, P. Lombardo, and C. J. Baker, "A Multistage Processing Algorithm for Disturbance Removal and Target Detection in Passive Bistatic Radar," *IEEE Transactions on Aerospace and Electronic Systems*, vol. 45, pp. 698–722, 4 2009.

[40] F. S. V. Bazán, "CGLS-GCV: a hybrid algorithm for low-rank-deficient problems," *Applied Numerical Mathematics*, vol. 47, pp. 91–108, 2003.

[41] D. M. Pozar, *Microwave engineering.* Wiley, 1997.

[42] "Application Note IMD Measurements with IMDView ." Available at `https://dl.cdn-anritsu.com/en-us/test-measurement/files/Application-Notes/Application-Note/11410-00859B.pdf`.

[43] Republic of South Africa, "Government Gazette Staatskoerant Part 1 OF 8," vol. 635, pp. 41650–9.

[44] "Siretta Delta 7A Datasheet." Available at `https://www.siretta.co.uk/product.php?id=871`.

[45] "Altair Feko Overview." Available at `https://altairhyperworks.com/product/FEKO`.

[46] "HMC1122 Datasheet and Product Info — Analog Devices." Available at `https://www.analog.com/en/products/hmc1122.html`.

[47] "Printed Circuit Board Manufacturer — Cape Town — TraX Interconnect." Available at `https://www.trax.co.za/`.

[48] "Microwave Office Software - RF/Microwave Circuit Design — NI AWR Design Environment." Available at https://www.awrcorp.com/products/ni-awr-design-environment/microwave-office-software.

[49] "HMC649A Datasheet and Product Info — Analog Devices." Available at https://www.analog.com/en/products/hmc649a.html.

[50] D. M. P. Smith, D. B. Davidson, A. Bester, and J. Andriambeloson, "Modernising, upgrading and recommissioning the indoor antenna range at Stellenbosch University," Tech. Rep. 1, 2016.

[51] "Ettus Research - Product Detail." Available at https://www.ettus.com/product/details/UN210-KIT/.

[52] "Mini-Circuits ZX05-U742MH-S+." Available at https://www.minicircuits.com/WebStore/dashboard.html?model=ZX05-U742MH-S%2B.

[53] "Mini-Circuits ZX05-C60MH-S+." Available at https://www.minicircuits.com/WebStore/dashboard.html?model=ZX05-C60MH-S%2B.

[54] "Mini-Circuits ZX60-83LN-S+." Available at https://www.minicircuits.com/WebStore/dashboard.html?model=ZX60-83LN-S2B.

[55] "Mini-Circuits ZX60-V63+." Available at https://www.minicircuits.com/WebStore/dashboard.html?model=ZX60-V63%2B.

[56] "Agilent Spectrum Analyzer Measurements and Noise Application Note 1303 Measuring Noise and Noise-like Digital Communications Signals with a Spectrum Analyzer," tech. rep.

www.ingramcontent.com/pod-product-compliance
Lightning Source LLC
LaVergne TN
LVHW041230150826
845673LV00008B/2332

* 9 7 8 3 3 8 4 2 4 8 9 6 1 *